Praise for

## *Why Beautiful People Have More Daughters*

"A rollicking bit of pop science that turns the lens of evolutionary psychology on issues of the day." —Dan Neil, *Los Angeles Times*

"Even if some of their findings are startlingly politically incorrect, it's impossible to read this book and not see yourself throughout. You can't help it: It's in your genes." —*The Portland Mercury*

"Miller and Kanazawa provide a new take on the nature-versus-nurture debate." —*Science News*

"This accessible book opens the youthful field of evolutionary psychology wide for examination, with results often as disturbing as they are fascinating." —*Publishers Weekly*

"Answer[s] burning questions like why women tend to lust after males who already have mates . . . and why newborns look more like Daddy than Mommy." —*New York Post*

"Written as a series of provocative questions and answers (for example, 'Why Are Most Suicide Bombers Muslim?'), *Why Beautiful People Have More Daughters* is destined to inspire passionate debate—after all, arguing is just human nature." —VeryShortList.com

"An absolute gem of a book. [It] possesses that rare combination of being gripping from start to finish, combined with the latest scientific theories and findings on the things we humans hold most dear—sex, mating, family, status, and violence. I literally couldn't put it down." —David M. Buss, author of *Evolutionary Psychology*

*continued . . .*

"An exuberant, accessible, exhilarating, intellectually aerobic workout of an introduction to the new science of human nature."

—David P. Barash,
Professor of Psychology, University of Washington,
and author of *Madame Bovary's Ovaries*

"A first-rate introduction to evolutionary psychology. . . . Miller and Kanazawa shatter the myth that evolutionary psychology merely confirms intuition, showing with wit and verve how it can lead to surprising and counterintuitive conclusions."

—*Kingsley R. Browne,*
Professor, Wayne State University Law School,
and author of *Biology at Work*

# Why Beautiful People Have More Daughters

From Dating, Shopping, and Praying to Going to War and Becoming a Billionaire—Two Evolutionary Psychologists Explain Why We Do What We Do

**Alan S. Miller** and **Satoshi Kanazawa**

A Perigee Book

**A PERIGEE BOOK**
**Published by the Penguin Group**
**Penguin Group (USA) Inc.**
**375 Hudson Street, New York, New York 10014, USA**
Penguin Group (Canada), 90 Eglinton Avenue East, Suite 700, Toronto, Ontario M4P 2Y3, Canada (a division of Pearson Penguin Canada Inc.) • Penguin Books Ltd., 80 Strand, London WC2R 0RL, England • Penguin Group Ireland, 25 St. Stephen's Green, Dublin 2, Ireland (a division of Penguin Books Ltd.) • Penguin Group (Australia), 250 Camberwell Road, Camberwell, Victoria 3124, Australia (a division of Pearson Australia Group Pty. Ltd.) • Penguin Books India Pvt. Ltd., 11 Community Centre, Panchsheel Park, New Delhi—110 017, India • Penguin Group (NZ), 67 Apollo Drive, Rosedale, North Shore 0632, New Zealand (a division of Pearson New Zealand Ltd.) • Penguin Books (South Africa) (Pty.) Ltd., 24 Sturdee Avenue, Rosebank, Johannesburg 2196, South Africa

Penguin Books Ltd., Registered Offices: 80 Strand, London WC2R 0RL, England


Cover art by CSA Images
Cover design by Ben Gibson



PRINTING HISTORY
Perigee hardcover edition / September 2007
Perigee trade paperback edition / September 2008

Perigee trade paperback edition ISBN: 978-0-399-53453-9

The Library of Congress has cataloged the Perigee hardcover edition as follows:

Miller, Alan S.
Why beautiful people have more daughters: from dating, shopping, and praying to going to war and becoming a billionaire—two evolutionary psychologists explain why we do what we do / Alan S. Miller and Satoshi Kanazawa.— 1st ed.
p. cm.
"A Perigee book."
Includes bibliographical references and index.
ISBN 978-0-399-53365-5 (alk. paper)
1. Evolutionary psychology. 2. Beauty, Personal—Psychological aspects. 3. Sex differences (Psychology) I. Kanazawa, Satoshi. II. Title.
BF698.95.M545 2007
155.7—dc22 2007011491

PRINTED IN THE UNITED STATES OF AMERICA

10 9 8 7 6 5 4 3 2 1

To our long-suffering foreign wives:

平峰　美代子

and

Заливина Ирина Владимировна

A. S. M. & S. K.

## Contents

# Preface

I first met Alan S. Miller in the Department of Sociology at the University of Washington, where I began my graduate study in 1985. Alan joined me in the department a year later, after having received his master's degree from California State University, Dominguez Hills, in 1986. Both Alan and I were trained in a field of sociology called rational choice theory, an application of microeconomic theory to sociological problems. After receiving my master's degree from Washington, I moved to the University of Arizona to pursue my PhD. Alan received his PhD from Washington in 1991, and took up his first teaching post at the University of North Carolina at Charlotte.

Years went by, and Alan and I kept in touch by email and telephone, although the very last time I ever saw Alan in person was August 1993 in Miami Beach, during the annual conference of the American Sociological Association. In 1998, Alan was beginning to write a book on group conformity and social order in Japan from a rational choice perspective. He had been teaching a course on the topic at Hokkaido University, to which he moved in 1996, and was frustrated by the lack of good textbooks in the subject area, so he decided to write one himself. He could write all of the empirical chapters on various aspects of life in Japan by himself, being an established Japan specialist and having lived in the country for a few years. But he needed help with writing the theoretical chapters, and asked me if I wanted to write them as the second author of the book. It was a very generous offer; he gave me joint authorship of the book for writing only two chapters in it. So even though I knew very little about Japan, we decided to write the book together. It was published in 2000.[1]

When Alan and I began collaborating on our first book, I started pitching evolutionary psychology to him. Alan was "hooked" instantaneously. He realized its tremendous value, as I did, and started reading evolutionary psychology voraciously. We later said that the best thing that came out of our collaboration on our first book was not the book itself but Alan's conversion to evolutionary psychology. We could never stop talking about it; it is just that good. It is an endless fountain of ideas.

In September 2000, soon after our first book was published, Alan had an idea for our second book. He thought it would be great for us to write an introduction to evolutionary psychology for a general nonacademic audience. He also had the idea to make it an adult version of a children's question-and-answer book of science. I thought it was a fantastic idea, so we began collaborating on our second book immediately.

Then, in early 2001, Alan fell ill and was later diagnosed with Hodgkin's disease. He spent two years in and out of hospital, undergoing several major operations and constant chemotherapy and radiation therapy. Despite the fact that his initial prognosis was reasonably good (70 percent chance of survival), he became progressively ill. I visited Sapporo in late December 2002, but I was not able to see him; he was simply too ill, and the doctors didn't allow any visitors except for his wife. He died on January 17, 2003, at the tragically young age of 44. He is the closest friend whom I have ever lost to date, and I am not sure if I can ever get over it.

Before his death, while he was still relatively healthy, Alan was able to complete the first draft of several chapters of this book. He had also seen and commented on the first draft of several other chapters that I had written. However, as the nature of the book has changed since his death, I have had to rework all of Alan's chapters, while retaining his original ideas.

Thus, Alan never had a chance to see the final manuscript or approve the subsequent revisions that I made to his chapters. I am keeping Alan as the first author of this book because that was the arrangement we agreed upon when we began our collaboration, and because the book was originally his idea. However, the reader should know that I am solely responsible for the entire contents of the final manuscript, which Alan did not have a chance to see or approve. Alan should be credited as the genius behind this book, while any remaining shortcomings should be attributed to me.

–Satoshi Kanazawa

# Acknowledgments

This book has been a long time in the making. Due to the controversial and often politically incorrect nature of its subject matter, we have encountered obstacles and resistance from many corners in our effort to publish this book.

For bringing this project to long-delayed fruition, I must first of all thank my editor at Perigee, Marian Lizzi, for believing in me and in this project as no other editors have. Until I met Marian, I had not known that editors could be so supportive and encouraging; I had never had an editor who I felt was completely "on my side" and had my interests at heart. I could not have found Marian without my able literary agent, Andrew Stuart, who, through his worldwide network of subagents, also found several foreign publishers who wanted to translate and publish this book on three other continents.

I thank Laura L. Betzig, Martie G. Haselton, and Kaja Perina for commenting on parts of the book in earlier forms, and Sarah E. Hill, Bobbi S. Low, Frank J. Sulloway, and Robert L. Trivers for providing me with various pieces of information. I must apologize to David M. Buss because, even though he has been supportive of me and this book project, I know he will not like its title. (But it was Marian's idea!)

Two editors in chief of *Psychology Today* have played significant roles in the course of both my career and the development of this book project. Hara Estroff Marano, former editor in chief of *PT* and current editor at large, was the very first journalist ever to interview me and feature my scientific work in the media. We have since become friends, and she has been very supportive of me throughout the years, offering me sage advice, warm support, and delicious ginger-poached pears. Most importantly, she introduced me to her agent, Andrew Stuart; and to her colleague Kaja Perina.

My dear friend Kaja Perina, current editor in chief of *PT*, has been a truly unfailing supporter of me, my career, and this book project. Kaja is the most frighteningly intelligent, cultured, sophisticated, and beautiful woman that I know. She became editor in chief of *PT* at the astonishingly young age of 27. At the current pace, Kaja can become President of the United States on her 35th birthday, except that it would be far beneath her to take on such a job. Despite her enormously important and demanding job, Kaja is always available to talk whenever I make trans-Atlantic phone calls to her desk in New York, ready with brilliant insight and judicious advice. Kaja has been the wisdom and maturity behind this book, since I have neither.

While working on this book, I made the transition from untenured Assistant Professor on an annual contract at a small college in Western Pennsylvania, to tenured Reader on a lifetime contract at the London School of Economics and Political Science, among other things. For never losing faith in me during the years I was in academic exile, I thank Denise L. Anthony, Paula England, Roberto Franzosi, Debra Friedman, Michael Hechter, Christine Horne, Lisa A. Keister, Michael W. Macy, Toshio Yamagishi, Lawrence A. Young, and–above all and always–Mary C. Still. Thanks and unbounded love to Cary Lee Coryell and Abigail Iris Coryell for being the greatest daughters that a delusional man can pretend to have. I am grateful to Bruce J. Ellis, Diane J. Reyniers, and David de Meza for rescuing me.

–Satoshi Kanazawa

## Introduction

# Human Nature "Я" Us

This book is about human nature. "Human nature" is one of those things that everybody knows and uses in their daily conversation, but that is difficult to define precisely. What *is* human nature?

The answer is both complex and remarkably simple. Every time we fall in love, every time we fight with our spouse, every time we enjoy watching our favorite TV show, every time we get scared walking at night in bad neighborhoods where tough young men loiter, every time we are upset about the influx of immigrants into our country, every time we go to church, we are–*in part*–behaving as a human animal with its own unique evolved nature–human nature.

This means two things. First, our thoughts, feelings, and behavior are produced not only by our individual experiences and environment in our own lifetime, but also by what happened to our ancestors millions of years ago. Our human nature is the cumulative product of the experiences of our ancestors in the past, and it affects how we think, feel, and behave today.

Second, because human nature is universal–sometimes shared by all humans, sometimes only shared by members of our own sex–our thoughts, feelings, and behavior are shared, to a large extent, by all other humans on earth (or all other men or women). Despite the seemingly large cultural differences in various societies, our daily experiences are essentially the same as those of people from Aberdeen, Bombay, and Cairo, to Xian, Yukon, and Zanzibar.

Human behavior is a product *both* of our innate human nature *and* of our unique individual experiences and environment. Both are important influences on our thoughts, feelings, and behavior. In this book, we emphasize human nature to the near exclusion of experience and environment. But there is a very good reason for that.

## The Forgotten Half of the Equation

Everyone agrees that experience and environment are both important influences on human behavior. Despite critics' claims to the contrary, there are no serious biological or genetic determinists in science.[1] There are a few genetic diseases, such as Huntington's disease, which are 100 percent determined by genes; if someone carries the affected gene, they will develop the disease no matter what their experiences or environment.[2] An individual's eye color and blood type are also 100 percent determined by genes. So these (and a few other) traits are entirely genetically determined. Otherwise, there are no human traits that are 100 percent determined by genes. Nor are there any serious scientists who think there are.

However, there are many social scientists, journalists, and others who believe that human traits and behavior are almost entirely determined by the environment.[3] As we will see in chapters 1 and 2, most social scientists tend to be *environmental determinists.* They believe that individual experiences and social environments *completely* determine human behavior, and there are no roles played by genetic and biological factors.

We emphasize biological influences on human behavior not because they are more important than environmental influences but because they need to be emphasized, for while no human behavior is 100 percent determined by the genes, *neither is any human behavior 100 percent determined by the environment.* The former is not controversial; everybody knows it. The latter is controversial, and not enough people know it. That is why we emphasize it in this book.

Evolutionary psychology is the new science of human nature and, together with behavior genetics, is currently the best theoretical perspective with which to understand the biological and evolutionary influences on human preferences, values, emotions, cognition, and behavior.[4] In this book, we try to introduce evolutionary psychology to a wider audience. While evolutionary psychology is sweeping the social and behavioral sciences, there have been relatively few *recent* popular introductions to the field.[5] Because there are a large number of fascinating studies published every year in evolutionary psychology, popular introductions must be updated every so often.

In doing so, we adopt a question-and-answer format. We try to use evolutionary psychology to address and explain typical experiences in our daily lives as well as events and problems in the larger society, because we believe insight from evolutionary psychology can shed new light on and provide novel solutions to some old problems. This is, if you will, an evolutionary psychology question-and-answer book.

We also want to demonstrate in this book that evolutionary psychology is not just about sex and mating. While there have been many fascinating studies in evolutionary psychology about sex and human mating (and we will discuss them in chapter 3), we believe evolutionary psychology can explain much more about human behavior. In fact, we want to show that insight from evolutionary psychology is useful in explaining puzzles in *all* areas of human life. This is the reason we adopt the question-and-answer format and address many areas of human social life in this book.

## Two Errors in Thinking That We Must Avoid

In any discussion of evolutionary psychology, it is very important to avoid two serious mistakes in thinking. They are called the naturalistic fallacy and the moralistic fallacy. The *naturalistic fallacy*, which was coined by the English philosopher George Edward Moore in the early twentieth century[6] though first identified much earlier by the Scottish philosopher David Hume,[7] is the leap from *is* to *ought*—that is, the tendency to believe that what is natural is good; that what is, ought to be. For example, one might commit the error of the naturalistic fallacy and say, "Because people *are* genetically different and endowed with different innate abilities and talents, they *ought* to be treated differently."

The *moralistic fallacy*, coined by the Harvard microbiologist Bernard Davis in the 1970s,[8] is the opposite of the naturalistic fallacy. It refers to the leap from *ought* to *is*, the claim that the way things should be is the way they are. This is the tendency to believe that what is good is natural; that what ought to be, is. For example, one might commit the error of the moralistic fallacy and say, "Because everybody *ought* to be treated equally, there *are* no innate genetic differences between people." The science writer Matt Ridley calls it *reverse naturalistic fallacy.*[9]

Both are errors in thinking, and they get in the way of progress in science in general, and in evolutionary psychology in particular. However, as Ridley astutely points out, political conservatives are more likely to commit the naturalistic fallacy ("Nature designed men to be competitive and women to be nurturing, so women ought to stay home to take care of their children and leave politics to men"), while political liberals are equally likely to commit the moralistic fallacy ("The Western liberal democratic principles hold that men and women ought to be treated equally, and therefore men and women are biologically identical and any study that demonstrates otherwise is *a priori* false"). Since academics, and social scientists in particular, are generally left-leaning liberals, the moralistic fallacy has been a much greater problem in academic discussions of evolutionary psychology than the naturalistic fallacy. Most academics are above committing the naturalistic fallacy, but they are not above committing the moralistic fallacy.

We will avoid both errors–both leaps of logic–in this book by never talking about what ought to be at all and only talking about what is. It is not possible to make either mistake if we never talk about *ought.* We will not draw moral conclusions from the empirical observations we describe in subsequent chapters, and we will not be guided in our observations by moral principles.

There are only two legitimate criteria by which you may evaluate scientific ideas and theories: logic and evidence. Accordingly, you may justifiably criticize evolutionary psychological theories (including those presented in this book) if they are logically inconsistent within themselves or if there is credible scientific evidence against them. As scientists, we will take all such criticisms seriously. However, it would hardly be appropriate to criticize scientific theories simply because their implications are immoral, ugly, contrary to our ideals, or offensive to some. We can tell you right now that the implications of many of the ideas we present in this book (whether ours or someone else's) are indeed immoral, ugly, contrary to our ideals, or offensive to either men or women (or some other groups of people). However, we must state them as they are because, to the best of our scientific judgment, they are true. That does not mean that we endorse all possible consequences and implications of our observations or believe that they are somehow good, right, desirable, or justifiable.

Truth is the guiding principle in science, and it is the most important thing for scientists. We also believe that any solution to a social problem must start with the correct assessment of the problem itself and its possible causes. We can never devise a correct solution to a problem if we don't know what its ultimate causes are. So the true observations are important foundations of both basic science and social policy.

## A Note about Stereotypes

It would be tempting to dismiss many of our observations (such as answers to questions like, "Why are there so many deadbeat dads but so few deadbeat moms?" or "Why are almost all violent criminals men?") as stereotypes. We plead guilty to the charge; many of our (and others') observations *are* stereotypes. But we suggest that you cannot dismiss an observation by calling it a stereotype, as if that suddenly makes it untrue and thus unworthy of discussion and explanation. In fact, the opposite is the case. *Many stereotypes are empirical generalizations with a statistical basis and thus on average tend to be true.* The only problem with stereotypes and empirical generalizations is that they are not always true for all individual cases.

There are always individual exceptions to stereotypes. There are many dedicated fathers and female criminals, even though the generalizations are still true. The danger lies in applying the statistical generalizations to individual cases, which may or may not be exceptions.

Stereotypes have a bad name, but many of them may turn out to be true empirical generalizations that someone does not like or that are unkind or offensive to some groups. An observation, if true, becomes an empirical generalization until someone objects to it, and then it becomes a stereotype. For example, the statement "Men are taller than women" is an empirical generalization. It is in general true (and, by the way, there are evolutionary psychological explanations for this phenomenon[10]), but there are individual exceptions. There are many men who are shorter than the average woman, and there are many women who are taller than the average man, but these exceptions do not make the generalization untrue; in every human society, men on average are taller than women. Everybody knows this, but nobody calls it a stereotype because it is not unkind to anybody. Men in general like being taller than women, and women in general like being shorter than men.[11]

However, as soon as one turns this around and makes the slightly different, yet equally true, observation that "Women are fatter than men," it becomes a stereotype because nobody, least of all women, wants to be considered fat. But it is true nonetheless; women have a higher percentage of body fat than men throughout the life course (and there are evolutionary reasons why this is the case as well[12]). Once again, there are numerous individual exceptions, but the generalization still holds at the population level.

In this book, we will attempt to make and then explain observations and phenomena that the available scientific evidence indicates are empirically true, even though there are individual exceptions and *regardless of whether they may seem unkind to some groups* (which they may in many cases). We draw no consequences or conclusions out of such observations; we are simply stating and explaining them. We will not commit either naturalistic or moralistic fallacy. Stereotypes and empirical generalizations are neither good nor bad, desirable nor undesirable, moral nor immoral. They just are.

Stereotypes also do not tell us how to behave or treat other people (or groups of people). Stereotypes are observations about the empirical world, not behavioral prescriptions. One may not infer how to treat people from empirical observations about them. Stereotypes tell us what groups of people tend to be or do in general; they do not tell us how we ought to treat them. Once again, there is no place for "ought" in science.

## How to Use This Book

The book is organized so that after the introduction and two introductory chapters, readers may skip around and read whichever chapters and sections are of interest. Each chapter (and section within it) is designed to be self-contained for anyone who has read the introduction and the first two chapters. We introduce fundamental principles of evolutionary psychology in chapters 1 and 2. Chapters 3–8 cover different areas of everyday life (sex and mating, marriage, family, crime and violence, political and economic inequalities, and religion and conflict). In each we pose several questions readers may have wondered about in their own lives, and provide evolutionary psychological answers to them. We look ahead at the questions that still remain unanswered by evolutionary psychology in the conclusion.

All of our claims are fully referenced to scientific studies that provide supportive evidence. For endnotes that simply give citation information, we use the standard endnote reference numbers, for example.[1] References that give greater information not contained in the text have reference numbers with brackets, for example.[2]

# 1

# What Is Evolutionary Psychology?

Evolutionary psychology is a new, emerging field. The first landmark studies in evolutionary psychology were published in the late 1980s,[1] and the birth of modern evolutionary psychology was marked in 1992 with the publication of the tome *The Adapted Mind: Evolutionary Psychology and the Generation of Culture*,[2] which is often regarded as the bible of modern evolutionary psychology.[3] What was there before then? Before we tackle the question, "What is evolutionary psychology?" in this chapter, let's pause a moment to consider what theories and explanations were available to social scientists before its advent.

## A Typical View from the Social Sciences

Most social scientists explain human behavior in a more or less typical fashion. The particular school of thought is called "the Standard

Social Science Model."[4] Because social scientists and their theories tend to have a lot of influence on the general public, the same view also characterizes how ordinary people account for human behavior in their everyday lives.

What exactly is the Standard Social Science Model? A set of related principles characterize its main tenets.

1. *Humans are exempt from biology.* Social scientists who subscribe to the Standard Social Science Model know that biology (and its branches, like zoology, ornithology, and entomology) can explain the behavior of all other species in nature. Yet they make an exception for humans as a sole species whose behavior is not explained by biological principles and theories. Human exceptionalism is the hallmark of the Standard Social Science Model. Many social scientists have averse reactions to biological explanations of human behavior.[5] This principle says that humans are exceptions in nature.

2. *Evolution stops at the neck.*[6] Social scientists in the Standard Social Science Model tradition, who do not believe in biological influences on human behavior and cognition, nonetheless acknowledge that human anatomy has been shaped by evolution. They recognize that human body parts, such as the fingers and the toes, are the way they are because of a long evolutionary process of natural and sexual selection. However, they contend that evolution has had no effect on the contents of the human brain and the human mind. This principle says that the brain is an exception in the human body.

3. *Human nature is tabula rasa (a blank slate).*[7] As a result of principle 2 above, social scientists in the Standard Social Science Model

tradition contend that humans are born with a mind like a blank slate. Once again, they recognize that all the other species have innate natures: dogs have an innate dog nature, which makes them behave more or less the same no matter where they live or what their individual life experiences have been, and cats have an innate cat nature, which similarly makes them behave the same but different from dogs. The same goes for all species in nature–*except for humans.* Humans do not have an innate nature, as they are born with minds that are blank slates. Principle 3, like principle 1, is an example of human exceptionalism.

4. *Human behavior is a product almost entirely of environment and socialization.* Since, according to the Standard Social Science Model, humans have no innate human nature that guides their behavior, the contents of human nature must be written after birth. The Standard Social Science Model contends that this occurs by a lifelong process of socialization (learning via instruction, imitation, copying, etc.) by the agents of socialization (parents, other family members, teachers, other adults in society, the media). Humans become the way they are because of socialization; socialization makes them human. In particular, men and women acquire their typical male and female behavior through gender socialization. This is why another name for the Standard Social Science Model is *environmentalism.* Most social scientists believe that the environment and life experiences almost entirely shape and determine human behavior.

Admittedly, this is a somewhat simplified version of the Standard Social Science Model, but it is not far off the mark. Not all social scientists agree with all of the four tenets, but most would agree with most of them to a large extent (and many agree with all).[8] Recent surveys of introductory textbooks in sociology and psychology reveal very cursory (and often incorrect) discussions of human evolution and its effect on behavior.[9]

## The Evolutionary Psychological Perspective[10]

Let us now look at the basic principles of evolutionary psychology. You could not possibly miss the sharp contrast between the Standard Social Science Model and evolutionary psychology.

As we said in the introduction, evolutionary psychology is the study of human nature. While the phrase "human nature" is used in common discourse to mean something essential but otherwise undefined about being human, it has a specific meaning in evolutionary psychology. It refers to a collection of components called *evolved psychological mechanisms* or *psychological adaptations* (these two terms are roughly synonymous). Human nature is the sum of such evolved psychological mechanisms, and evolutionary psychologists aim to discover more and more such psychological adaptations in humans. What, then, is an evolved psychological mechanism or psychological adaptation?

### *Evolved Psychological Mechanisms: What Human Nature Is Made Of*

An adaptation is a product of evolution by natural and sexual selection,[11] and it allows an organism to solve particular problems.[12] Our body is full of adaptations. Our eye is an adaptation; it allows us to see, navigate efficiently and safely, find prey, and avoid predators. Our hand is an adaptation; it allows us to hold and manipulate objects efficiently, collect and eat food, throw objects, and use and manufacture tools. If you imagine what your life would be like without an eye or a hand, you can begin to see the range of problems that these physical adaptations solve. Problems that adaptations allow us to solve are called adaptive problems. Adaptive problems are problems of survival and reproduction. Without solving adaptive problems, we will not be able to live long or reproduce successfully.

Psychological adaptations are like these physical adaptations in our body, except they are in our brain. They allow us to solve some adaptive problems by predisposing or inclining us to think or feel in certain ways. Just like we see or manipulate objects without much conscious thought, psychological adaptations often operate behind and beneath our conscious thinking. All adaptations (physical and psychological) are also *domain-specific*; they operate and solve problems only within a narrow area of life. The eye allows us to see but not manipulate objects; the hand allows us to manipulate objects but not see them. What the eye can do, the hand cannot, and vice versa. This is true for evolved psychological mechanisms as well; they only operate and solve problems in a narrow range of life.

Our preference for sweets and fats is an example of an evolved psychological mechanism.[13] Throughout most of human evolutionary history, getting enough calories was a serious problem; malnutrition and starvation were common. In this environment, those who, for reasons of random genetic mutation, had a "taste" for sweets and fats, which contain higher calories, were better off physically than those who did not have such a taste. Those who had a sweet tooth therefore lived longer, led healthier lives, and produced more healthy offspring than those who did not. They in turn passed on their (genetically influenced) taste to their offspring, over many thousands of generations. In every generation, those with this taste out-reproduced those without it, generation after generation, until most of us living today have a strong preference for sweets and fats.

Male sexual jealousy is another example of an evolved psychological mechanism.[14] Because gestation in humans and most other mammalian species occurs inside the female body, males of these species (including men) can never be certain that they are the father of their mates' offspring, while females are always certain of their maternity. In other words, the possibility of unwittingly raising children who are not genetically their own exists only for men. The technical term for this is *cuckoldry*. A man is *cuckolded* when his wife has an affair with someone, has a child by the lover, but successfully passes the child off as the husband's. According to one estimate, about 13–20 percent of children in the contemporary United States and 9–17 percent in contemporary Germany are not the genetic offspring of the man whose name appears on the child's birth certificate.[15] Another study shows that about 10–14 percent of children in Mexico have legal fathers different from their genetic fathers.[16] Earlier estimates from the US, the UK, and France range around 10–30 percent of all children.[17] As anyone who's ever watched a daytime talk show knows, concerns about biological paternity are far from a remote theoretical possibility; in fact, anywhere from one out of ten to one out of three children are raised by men who are unrelated to them genetically.

In evolutionary terms, men who are cuckolded and invest their financial and emotional resources in the offspring of other men end up wasting these resources, as their genes will not be represented in the next generation. For this reason, men have a strong evolutionary reason to be sexually jealous, while women, whose maternity is always certain, do not. The same psychological mechanism of sexual jealousy often leads to men's attempts to guard their mates physically, in order to minimize the possibility of their mates' sexual contact with other men, sometimes with tragic consequences.[18]

While men and women present the same frequency and intensity of their jealousy in romantic relationships,[19] there are clear sex differences in what triggers jealousy. The evidence from surveys and from physiological studies conducted in different cultures indicates that men become jealous of their mates' *sexual infidelity* with other men, underlying their reproductive concern for cuckoldry. In contrast, women become jealous of their mates' *emotional involvement* with other women, because emotional involvement often leads to diversion of their mates' resources from them and their children to their romantic rivals.[20] While recent critics of evolutionary psychology have questioned these conclusions mostly on methodological grounds,[21] both strong evolutionary logic and a preponderance of empirical evidence support the clear sex differences in romantic jealousy described above.[22]

### *Hardwired, Not Hardheaded*

Recall that evolved psychological mechanisms mostly operate behind and beneath conscious thinking. We do not consciously *choose* or *decide* to like sweets and fats. We like them but we do not know why; sweet and fatty foods just taste good to us. Similarly, we do not consciously *choose* or *decide* to feel jealous. We feel jealous under some circumstances, in response to certain predictable triggers, but we do not always know why. Evolutionary psychology contends that these evolved psychological mechanisms are behind most of our preferences, desires, and emotions, and they incline us to behave in certain ways. Evolutionary psychology explains human behavior in terms of the *interaction* between these evolved psychological mechanisms; the preferences, desires, and emotions that they produce in us; *and* the current environment in which they express themselves.

This is why both biology and environment are important components of any complete explanation for human behavior, even though, for reasons we noted in the introduction, we tend to emphasize the biological factors more in this book.

Evolutionary psychology is an application of evolutionary biology to human behavior. It is characterized by the following four principles, which form very clear contrasts to the four principles of the Standard Social Science Model, which we discussed above.

1. *People are animals.*[23] The first and most fundamental principle of evolutionary psychology is that there is nothing special about humans. They are just like all the other animal species. Now that does *not* mean that humans are not unique; they are. But then so are all other species. If humans are not unique, they would not be a separate species. The reason why human beings are a separate species is because no other species have exactly the set of characteristics that humans do. But the same thing can be said of chimpanzees, gorillas, dogs, cats, and giraffes. Humans are unique, but no more or no less so than fruit flies. Evolutionary psychology recognizes that the same biological laws of evolution apply to humans as they do to all other species. It therefore refutes the human exceptionalism of the Standard Social Science Model. In the words of the great sociobiologist Pierre L. van den Berghe, "certainly we are unique, but we are not unique in being unique. Every species is unique and *evolved* its uniqueness in adaptation to its environment."[24]

2. *There is nothing special about the human brain.* For evolutionary psychologists, the brain is just another body part, just like the hand or the pancreas. Just as millions of years of evolution have gradually shaped the hand or the pancreas to perform certain functions, so has evolution shaped the human brain to perform its function,

which is solving adaptive problems to help humans survive and reproduce successfully. Evolutionary psychologists apply the same laws of evolution to the human brain as they do to any other part of the human body. Evolution does *not* stop at the neck; it goes all the way up.

3. *Human nature is innate.* Just as dogs are born with innate dog nature, and cats are born with innate cat nature, humans are born with innate human nature. This follows from principle 1 above. What is true of dogs and cats must also be true of humans. Socialization and learning are very important for humans, but humans are born with the capacity for cultural learning, which is innate. Culture and learning are part of the evolutionary design for humans. Socialization merely reiterates and reinforces what is already in our brain (like the sense of right and wrong). This principle of evolutionary psychology is in clear contrast to the blank slate ("tabula rasa") assumption of the Standard Social Science Model. In the memorable words of William D. Hamilton, who is often regarded as the greatest Darwinian since Darwin, "The *tabula* of human nature was never *rasa* and it is now being read."[25] Evolutionary psychology is devoted to reading the tabula of human nature.

4. *Human behavior is the product of both innate human nature and the environment.* Genes very seldom express themselves in a vacuum. Their expressions–how the genes translate into behavior–often depend on and are guided by the environment. The same genes can express themselves differently depending on the context. In this sense, both innate human nature, which the genes program, and the environment in which humans grow up are equally important determinants of behavior. Unlike those in the school of the Standard Social Science Model, evolutionary psychologists do not

believe that human behavior is 100 percent determined by either factor. As we mentioned in the introduction, however, we will mostly focus on innate human nature, because this is the forgotten side of the equation.

## The Savanna Principle: Why Our Brains Are Stuck in the Stone Age

The second principle of evolutionary psychology discussed above—that there is nothing special about the human brain as a body part—leads to an important implication. Just as the basic shape and functions of the hand or the pancreas have not changed since the end of the Pleistocene Epoch ("the Ice Age") about ten thousand years ago, the basic functioning of the brain has not changed very much in the last ten thousand years. The human body (including the brain) evolved over millions of years in the African savanna and elsewhere on earth where humans lived during most of this time. This ancestral environment, where humans lived in small bands of 150 or so related individuals as hunter-gatherers, is called the environment of evolutionary adaptedness, or the ancestral environment.[26] It is to the ancestral environment that our body (including the brain) is adapted. Even though we live in the twenty-first century, we have a Stone Age brain (just like we have Stone Age hands and a Stone Age pancreas).

The evolved psychological mechanism produces adaptive behavior in the ancestral environment. Adaptive behavior is behavior that increases the chances of survival or reproductive success by solving the adaptive problems. Eating lots of sweet and fatty foods, which contain higher calories, is adaptive behavior that solves the adaptive problem of procuring sufficient food to survive. Becoming jealous at the remotest possibility of a mate's sexual infidelity, and guarding that mate so that she could not have sexual contact with other men, is adaptive behavior that solves men's adaptive problem of paternity uncertainty.

Our hominid ancestors spent 99.9 percent of their evolutionary history as hunter-gatherers on the African savanna and elsewhere on earth. It was not until about ten thousand years ago, when the Agricultural Revolution happened, that our ancestors started planting and cultivating their food through agriculture and animal husbandry. Almost everything we see around us today–cities, nation-states, houses, roads, governments, writing, contraception, TVs, telephones, and computers–came about in the last ten thousand years. Recall that our entire body is adapted to the ancestral environment and that we have a Stone Age body (including the brain). That means that our body is not necessarily adapted for things that came about since the end of the Pleistocene Epoch about ten thousand years ago. Ten thousand years is a very short period of time on the evolutionary time scale; it is simply not enough time for our body to make changes to accommodate things that came about in the meantime, especially since the environment has been changing too rapidly relative to how slowly we mature and reproduce. (It takes humans about twenty years to mature and be ready to reproduce. And, remember, only twenty years ago, for most people outside of the military and scientific circles, there was no such thing as the Internet or cell phones.) In other words, we still have the same evolved psychological mechanisms that our ancestors possessed more than ten thousand years ago.

This observation leads to a new proposition in evolutionary psychology called the Savanna Principle,[27] which states that

> *The human brain has difficulty comprehending and dealing with entities and situations that did not exist in the ancestral environment.*

One example of an entity that did not exist in the ancestral environment is TV or any other realistic images of other humans, such as photographs, videos, or films. The Savanna Principle would therefore predict that the human brain has difficulty comprehending and dealing with images shown on TV. This indeed appears to be the case.[28] A recent study shows that individuals who watch certain types of TV programs are more satisfied with their friendships, as if they had more friends or socialized with them more frequently. According to the Savanna Principle, this is probably because the human brain, adapted to the ancestral environment, has difficulty distinguishing between our real friends in the flesh and the characters we repeatedly see on TV. In the ancestral environment, any realistic images of other humans *were* other humans, and if you saw them repeatedly and they did not try to kill or harm you in any way, then more than likely they were your friends. Our Stone Age brain therefore assumes that the characters we repeatedly encounter on TV, very few of whom try to kill or harm us, are our real friends, and our satisfaction with friendships thereby increases by seeing them more frequently.

### *Maladaptive Adaptations*

Take the example of our preference for sweets and fats as an evolved psychological mechanism. This psychological mechanism solved the adaptive problem of survival in the ancestral environment by allowing those who possessed it to live longer. Our preferred consumption of sweets and fats was therefore adaptive *in the ancestral environment*. However, we now live in an environment where sweets and fats are abundantly available in every checkout line in every supermarket, in every city, in every industrial society, twenty-four hours a day, seven days a week. In other words, the original adaptive problem (malnutrition) no longer exists; very few people die of malnutrition in industrial societies. Yet we still possess the same psychological mechanism that compels us to consume sweets and fats. Because our environment is so vastly different from the ancestral environment, we now face a curious situation where those who behave according to the dictates of the evolved psychological mechanism are *worse off* in terms of survival. Obesity (to which overconsumption of sweets and fats leads) hinders survival. The Savanna Principle suggests that we continue to have (currently maladaptive) preferences for sweets and fats, and as a result become obese, because our brain cannot readily comprehend the supermarkets, the abundance of food in general, and indeed agriculture, none of which existed in the ancestral environment. Our brain still assumes we are hunter-gatherers with very precarious and unpredictable sources of food. If our brain truly comprehended supermarkets, we would not crave sweet and fatty foods.

Similarly, male sexual jealousy is another evolved psychological mechanism that hasn't quite caught up to modern times. It solved the adaptive problem of reproduction in the ancestral environment by allowing men who possessed it to maximize paternity certainty and minimize the possibility of cuckoldry. Sexual jealousy was therefore adaptive *in the ancestral environment.* However, sex and reproduction are often separated in the modern environment; many episodes of sex do not lead to reproduction. There is an abundance of reliable methods of birth control in industrial societies, and many women use the contraceptive pill. For these women, sexual infidelity does not lead to childbirth, and their mates will not have to waste their resources on someone else's children. Even if their mates cheated on them and got pregnant as a result, reliable paternity testing removes any paternity uncertainty. In other words, the original adaptive problem (paternity uncertainty) is less of a threat to reproductive success; men today are much less likely to invest unwittingly in someone else's genetic children. Yet men still possess the same psychological mechanism that makes them jealous at the possibility of their mates' sexual infidelity and compels them to guard their mates to minimize the possibility of cuckoldry. *The fact that his adulterous wife was on the Pill at the time of her sexual infidelity offers very little consolation to a man.*

Further, once again because our current environment is so vastly different from the ancestral environment, we now face a curious situation where those who behave according to the dictates of the evolved psychological mechanism are often *worse off* in terms of reproductive success. Extreme forms of mate guarding, such as violence against mates or romantic rivals, are crimes in most industrial nations. Incarceration, and consequent physical separation from their mates, does everything to reduce the reproductive success of the men. Yet men continue to exhibit sexual jealousy, and many men engage in extreme forms of mate guarding and vigilance, including violence.[29] The Savanna Principle suggests that this is because their brains cannot truly comprehend effective birth control, written laws, the police, and the courts. If they did, they would not engage in extreme forms of mate guarding (such as violence) or any other criminal behavior for which they would likely go to jail.

We caution you that the Savanna Principle as stated above was proposed very recently (even though it is based on observations made earlier by pioneers of evolutionary psychology)[30] and is not yet part of the established literature of evolutionary psychology. Its implications have yet to be subjected to rigorous experimental testing. However, we refer to it throughout the rest of the book, because we believe there is a kernel of truth to it and that it can explain a wide range of otherwise puzzling instances of human behavior.

### *Human Evolution Pretty Much Stopped about Ten Thousand Years Ago*

The Savanna Principle points to a couple of very important–but often neglected–observations about human evolution: Evolution happens very gradually, and natural selection requires a stable, unchanging environment to which it can respond.

Evolution takes many *generations*, and so the speed of evolution of a species is relative to how long it takes for individuals of the species to mature sexually. Evolution happens faster for fast-maturing species and slower for slow-maturing species. Fruit flies are one of the fastest-maturing species in nature, and humans are one of the slowest. It takes only seven days for fruit flies to mature sexually under ideal conditions, whereas it takes fifteen to twenty years for humans. It means that there can be more than fifty generations of fruit flies in one year, before a human baby can even begin to walk. There are more than a thousand generations of fruit flies in one human generation (twenty years), for which humans need more than twenty thousand years. Evolution for fruit flies can happen pretty fast, which is precisely the reason why they are the favorite species for geneticists to study. Human evolution happens much, much more slowly. No human scientists can see it in action the way they can observe fruit fly evolution unfold in the lab.

The second point is even more important: Natural selection under most circumstances requires a stable, unchanging environment for many, many generations. For example, if the climate is very cold for centuries and millennia, then gradually individuals who have better resistance to cold will be favored by natural selection, and their neighbors who have less resistance to cold (who are more adapted to hot climates) will die out before they can leave many children. This will happen generation after generation, until one day all humans have great resistance to cold. A new trait–resistance to cold–has now evolved and become part of universal human nature. But this trait could not have evolved if the climate was cold for one century (only five human generations, albeit 5,200 fruit fly generations) and then hot for another century, only to be cold again in the third century. Natural selection would not know who (with which traits) to select.

Since the advent of agriculture about ten thousand years ago and the birth of human civilization which followed, humans have not had a stable environment against which natural selection can operate. For example, a mere two centuries (ten generations) ago, the United States and the rest of the Western world were largely agrarian; most people were farmers. In the agrarian society, men achieved higher status by being the best farmers; those who possessed certain traits that made them good farmers had higher status and thus greater reproductive success than others who didn't possess such traits.

Then, only a century later, the United States and Europe were predominantly industrial societies; most men made their living working for factories. Traits that make men good factory workers (or, better yet, factory *owners*) may or may not be the same as the traits that make them good farmers. Certain traits–such as intelligence, diligence, and sociability–probably remain important,[31] but others–such as a feel for nature, the soil, and animals, and the ability to work outdoors or forecast weather–cease to be important, and other traits–such as punctuality, the ability to follow instructions, a feel for machinery or mechanical aptitude, and the ability to work *indoors*–suddenly become important.

Now we are in a post-industrial society, where most people work neither as farmers nor factory workers but in the service industry. Computers and other electronic devices become important, and an entirely new set of traits is necessary to be successful. Bill Gates and Sir Richard Branson (and other successful men of today) may not have made particularly successful farmers or factory workers. All of these dramatic changes happened within ten generations, and there is no telling what the next century will bring and what traits will be necessary to be successful in the twenty-second century. We live in an unstable, ever-changing environment, and have done so for about ten thousand years.

For hundreds of thousands of years before that, our ancestors lived as hunter-gatherers on the African savanna, in a stable, unchanging environment to which natural selection could respond. That is why all humans today have traits that would have made them good hunter-gatherers in Africa–men's great spatiovisual skills, which allowed them to follow animals on a hunting trip for days and for miles without a map or a global satellite positioning device and return home safely; and women's great object location memory, which allowed them to remember where fruit trees and bushes were and return there every season to harvest, once again without maps or permanent landmarks.

For the last ten thousand years or so, however, our environment has been changing too rapidly for evolution to catch up. Evolution cannot work against moving targets. That's why humans have not evolved in any predictable direction since about ten thousand years ago. We hasten to add that certain features of our environment have remained the same–we have always had to get along with other humans, and we have always had to find and keep our mates–so certain traits, like sociability or physical attractiveness, have always been favored by natural and sexual selection. But other features of our environment have changed too rapidly relative to our generation time, in a relatively random fashion–who could have predicted computers and the Internet a century ago?–so we have not been able to adapt and evolve against the constantly moving target of the environment.

# 2

# Why Are Men and Women So Different?

Much of our discussion in the following chapters hinges on differences between men and women. Now *everyone knows* that men and women are different. On the whole, they want different things, they are good at different things, and they behave in different ways. While everybody may know *that* men and women are different, they may not know *why*. Or they may think they do, but they might be wrong.

The prevailing explanation in the Standard Social Science Model, popular among academic social scientists and the general population alike, is *gender socialization.* According to this explanation, men and women (and boys and girls) think and behave differently because they have been socialized differently by their culture and society. Recall that the Standard Social Science Model contends that human nature is a blank slate (principle 3). Male and female babies are born identical except for a few anatomical differences, *but these anatomical differences do not include the brain* (principle 2). Since the day of their birth, boys and girls are treated differently and socialized either as boys or girls. Boys are encouraged to be aggressive and violent (by being given toy trucks and toy guns), while girls are taught to be caring and nurturing (by being given dolls and tea sets). Gender socialization permeates every aspect of culture and society (it is done not only by the parents but by educational, religious, political, and economic institutions and the media) and continues throughout the life course, and its effects are cumulative. By the time boys and girls grow up to be men and women, they think and behave differently because "society" expects them to, and the sex differences are apparently permanent. However, the Standard Social Science Model contends that if parents and "society" provide gender-neutral, androgynous socialization to children, then boys and girls will not behave differently, and men and women will be the same in their behavior, cognition, values, and preferences.

An overwhelming amount of evidence now available from science unambiguously demonstrates that this view is false. We will discuss only two recent studies here, and refer interested readers to more comprehensive reviews.[1]

### *Sex Differences Appear on the First Day of Life*

University of Cambridge psychologist Simon Baron-Cohen and his associates have conducted a careful experiment with one-day-old babies.[2] They simultaneously presented a picture of a woman's face and a mechanical mobile to 102 newborn babies (44 boys and 58 girls, but the researchers themselves were blind to the sex of these babies until after the experiment was finished). They videotaped the babies to measure which object they paid more attention to. Their analysis showed that more boys preferred to look at the mechanical mobiles, and boys on average gazed at them longer. In contrast, more girls preferred to look at the human face, and girls on average gazed at it longer. *Everybody knows* that boys and men tend to have greater interest in machines and other mechanical objects, and girls and women tend to be more social and express greater interest in relationships with others. If these sex differences are mostly the outcome of lifelong gender socialization, as the Standard Social Science Model claims, how can newborn babies who are just twenty-four hours old exhibit the same sex difference? Not even the most ardent supporters of the Standard Social Science Model would contend that twenty-four hours is enough for gender socialization.

### *Sex Differences Are Shared by Monkeys*

In a very ingenious experiment, Gerianne M. Alexander and Melissa Hines gave two stereotypically masculine toys (a ball and a police car), two stereotypically feminine toys (a soft doll and a cooking pot), and two neutral toys (a picture book and a stuffed dog) to 44 male and 44 female vervet monkeys.[3] They then assessed the monkeys' preference for each toy by measuring how much time they spent with each. Their statistical analysis demonstrated that male vervet monkeys showed significantly greater interest in the masculine toys, and the female vervet monkeys showed significantly greater interest in the feminine toys. The two sexes did not differ in their preference for the neutral toys. Alexander and Hines' article contains pictures of a female vervet monkey examining the genital area of the doll in an attempt to determine whether it is male or female, as a girl might, and of a male vervet monkey pushing the police car back and forth, as a boy might. If children's toy preferences were largely formed by gender socialization, as the Standard Social Science Model claims, in which their parents give "gender-appropriate" toys to boys and girls, how can these male and female vervet monkeys have the same preferences as boys and girls? They were never socialized by humans, and they had never seen these toys before in their lives.

As these two studies (and numerous others) show, the sex differences in behavior, cognition, values, and preferences are largely innate; universal across cultures; and, in many cases, constant across species.[4] If the sex differences were the result of social and cultural practices such as gender socialization, then they should by definition vary by culture and society. In fact, however, in every human society (and among many other species), males on average are more aggressive, violent, and competitive, and females on average are more social, caring, and nurturing. What is constant in every culture and society (sex differences in behavior) cannot be explained by what is variable across cultures and societies (cultural and social practices). A variable cannot explain a constant; only a constant can explain a constant.

## *A Consequence, Not a Cause*

Rather than the results of lifelong gender socialization, sex differences in behavior, cognition, values, and preferences are part of innate and distinct male and female human natures; men and women are hardwired to be different. Male and female human brains are different, just like male and female reproductive organs are different. Gender socialization helps to accentuate, solidify, perpetuate, and strengthen the innate differences between men and women, but it does not *cause* or *create* such differences. In other words, *men and women are not different because they are socialized differently; they are socialized differently because they are different.* Gender socialization is not the cause of sex differences; it is their consequence.

If gender socialization is not the cause of sex differences, then what is? What is the constant that explains the universal sex differences? It turns out that two simple biological facts lead to a whole array of sex differences: anisogamy and the internal gestation of fertilized eggs within the female body. *Anisogamy* means that the female sex cell (egg) is larger in size and fewer in number than the male sex cell (sperm). (This, by the way, is the biological definition of male and female. The female of any sexually reproducing species is defined as the sex that produces the larger sex cell, and the male, by default, is the other sex.) Anisogamy means that the egg is biologically far more valuable than the sperm; in nature, the sperm is abundant (practically infinite) in supply and biologically less costly to produce than the eggs. A quick rule of thumb in biology, which can explain a lot of sex differences in many species, is: *Sperm is cheap*.

The *internal gestation* of fertilized eggs within the female body means, among other things, that the female can produce far fewer offspring than the male can. It takes a woman at least nine months, usually a few years, to produce one child (because a woman is usually infertile while she nurses her baby); it takes a man fifteen minutes. A woman can normally have at most twenty to twenty-five pregnancies in her entire lifetime, usually far less; there is no limit to the number of children men can potentially produce. The operative term here, of course, is *potentially*.

Anisogamy and the internal gestation within the female body combine to produce a very important consequence: sex difference in fitness variance. *Fitness variance* is the difference between the "winners" and the "losers" in the reproductive game–how much more reproductively successful the winners are compared to the losers. Because of anisogamy and internal gestation, men have much greater fitness variance than women. Men's greater fitness variance means two things. First, looking at the bottom of the distribution, far more men remain childless than women, whereas relatively fewer women remain childless for life. So one consequence of greater fitness variance among men is that the *fitness floor* (the worst one can possibly do) is much lower for men than for women. The worst on average is much worse for men than for women.

Second, looking at the top of the distribution, a few men have a far larger number of children than any woman could possibly have. It is possible for some men to have dozens, hundreds, even thousands of children in their lifetimes, whereas a woman is limited to at most about twenty-five pregnancies in life. The other consequence of greater fitness variance among men is that the *fitness ceiling* (the best one can possibly do) is much higher for men than for women. The best is much better for men than for women. Fitness variance is the distance between the ceiling (the best) and the floor (the worst), so it is much greater for men than for women.

Even though anisogamy and the internal gestation within the female body makes greater fitness variance among men than among women *possible*, what actually produces it in reality is the fact that humans are naturally polygynous.[5] There is much confusion about terminology for different institutions of marriage, even among social scientists. *Monogamy* is the marriage of one man to one woman. *Polygyny* is the marriage of one man to more than one woman, while *polyandry* is the marriage of one woman to more than one man. *Polygamy* (although it is often used synonymously with polygyny in casual conversations) refers to both polygyny and polyandry. Because of its ambiguity, the word *polygamy* should not be used unless it specifically and simultaneously refers to both polygyny and polyandry.

Until very recently, humans were mildly polygynous throughout their evolutionary history.[6] Under polygyny, some men get more than their "fair share" of mates, leaving others with none. Thus, virtually all women, but not all men, get to reproduce, but those men who do, get to reproduce a large number of children. This is why few women, but relatively more men, have zero children (complete reproductive failure).

The largest number of children that a woman has ever had is sixty-nine. The wife of an eighteenth-century Russian peasant, Feodor Vassilyev, had twenty-seven pregnancies in her life, including sixteen pairs of twins, seven sets of triplets, and four sets of quadruplets; amazingly, Mrs. Vassilyev never had any single births in her life! And all but two of her sixty-nine children survived to adulthood. In contrast, the largest number of children that a man has ever had is *at least* 1,042.[7] The last Sharifian emperor of Morocco, Moulay Ismail the Bloodthirsty, maintained a large harem, as many ancient rulers did,[8] and had at least 700 sons and 342 daughters. (The exact number of children that Moulay Ismail had in his lifetime is lost to history, however, because they stopped counting them after a while.)

Exactly how many children Moulay Ismail the Bloodthirsty and Mrs. Feodor Vassilyev had is not important. What's important is this: The largest number of children that a man can potentially have is *two orders of magnitude greater* than the potential number of children that a woman can have (thousands vs. tens), while many men, but few women, face a great chance of ending their lives as total reproductive losers (leaving no offspring).

## Worth the Fight

As we will discuss repeatedly throughout this book, the greater fitness variance among men, rather than gender socialization, is the reason why men are much more aggressive, competitive, and violent than women. Men gain far more by competing with each other for access to mates, whereas the benefit of competition for women in reproductive terms is far less. If men compete successfully and gain reproductive access to a large number of women, they can potentially have hundreds, if not thousands, of children; if they fail to compete successfully, they face a distinct possibility of having no children at all. So the difference between a potential reward for competition and the potential cost of not competing is tremendous; they might as well compete. The same difference for women is much smaller. If women compete successfully and gain reproductive access to a large number of men, they can realistically have twenty to twenty-five children at most (in the absence of multiple births, which is beyond their control); if they fail to compete successfully, they might only have one or two children. The potential benefit of competition does not justify the potential costs (injury or death). This is why women are on the whole not as aggressive, competitive, or violent as men.[9]

The much higher fitness ceiling for men than for women also means that women make a far greater investment into their children than men do. While *reproductive success* is equally important for men and women (as it is for all living creatures), *each child* is far more important to a woman (as it is to females of all mammalian species) than it is to a man (as it is to males of all mammalian species). Each child represents a much greater portion of a woman's lifetime reproductive potential than it does a man's. It represents perhaps one-twentieth of a woman's lifetime reproductive potential; it represents one one-thousandth of a man's. Anisogamy and internal gestation thus lead to a large number of sex differences in behavior that we will discuss in subsequent chapters.

## *Exception That Proves the Rule*

One of the strongest pieces of evidence that anisogamy, the internal gestation, and the consequent greater fitness variance among men than among women lead to the sex differences in behavior comes from the proverbial "exception that proves the rule." While males have a greater fitness ceiling than females in most species, there are a few exceptional species for which this is not true. Among some fish, frog, and bird species, the males carry the fertilized eggs during gestation, and as a result, the females have a higher fitness ceiling than males do. Females of these species can continue reproducing while the males are "pregnant" with the young. As predicted by evolutionary biology, among these species, females are more aggressive, competitive, and violent than males.[10] Among these species, the females compete fiercely with each other for sexual access to the coy males. These exceptions therefore prove the rule that it is the fitness variance that determines which sex is more competitive and aggressive.[11]

## What about Culture? Is Anything Cultural?

Having read so far about how evolutionary psychology explains human behavior in terms of the interaction between evolved psychological mechanisms and the environment, you might be wondering, "Okay, that's fine and dandy. Our evolved mind does influence our behavior, as evolutionary psychologists say. But what about culture? Surely culture influences and molds human behavior through cultural socialization, as traditional sociologists say, even to a greater extent than our innate tendencies do."

Yes, culture and socialization do matter, to a certain extent. But the grave error of traditional sociologists and others under the influence of the Standard Social Science Model is to believe that human behavior is *infinitely* malleable, capable of being molded and shaped limitlessly in any way by cultural practices and socialization. Available evidence now shows that this view is false. Human behavior, while malleable, is not *infinitely* malleable by culture, because culture is not infinitely variable. In fact, despite all the surface and minor differences, evolutionary psychologists have shown all human cultures to be more or less the same.

### *There Is Only One Human Culture*

People–social scientists and laypersons alike–often speak of culture in the plural ("cultures") because they believe that there are many different cultures in the world. At one level, this is of course true; the American culture is different from the Chinese culture, both of which are different from the Egyptian culture, and so on. However, all the cultural differences are on the surface; deep down, at the most fundamental level, all human cultures are essentially the same.

To use a famous metaphor, coined by the cultural anthropologist Marvin Harris,[12] it is true that, at the surface level, people in some societies consume beef as food and worship pigs as sacred religious objects, while those in others consume pork as food and worship cows as sacred religious objects. So there is cultural variety at this concrete level. However, both beef and pork are animal proteins (as are dogs, whales, and monkeys), and both pigs and cows are animate objects (as are Buddha, Allah, and Jesus). And people in every human society consume animal proteins and worship animate objects. At this abstract level, there are no exceptions, and all human cultures are the same. There is no infinite variability in human culture, in the sense that there are no cultures in which people do not consume animal protein or worship animate objects.

To use another example, it is true that languages spoken in different cultures appear completely different, as anyone who ever tried to learn a foreign language knows. English is completely different from Chinese, neither of which is anything like Arabic. Despite these "surface" differences, however, all natural human languages share what the linguist Noam Chomsky calls the "deep structure" of grammar.[13] In this sense, English and Chinese are essentially the same, in the sense that beef and pork are essentially the same.

You need proof? Any developmentally normal child can grow up to speak any natural human language. Regardless of what language their genetic parents spoke, all developmentally normal children are capable of growing up to be native speakers of English, Chinese, Arabic, or any natural human language. In fact, when a group of children grow up together with no adults to teach them a language, they will invent their own natural human language with complete grammar. This does not mean, however, that the human capacity for language is infinitely malleable. Human children cannot grow up to speak non-natural languages like FORTRAN or symbolic logic, despite the fact that these are far more logical and easier to learn than any natural language (no irregular verbs, no exceptions to rules). Yes, a developmentally normal human child can grow up to speak any language, *as long as* the language is a product of human evolution, not a recent invention of computer scientists or logicians.[14]

Pierre van den Berghe, whom we encountered in the last chapter, again puts it best when he says, "Culture is the uniquely human way of adapting, but culture, too, evolved biologically."[15] Despite all the surface differences, there is only one human culture, because culture, like our body, is an adaptive product of human evolution. The human culture is a product of our genes, just like our hands and pancreas are.

Biologically, human beings are very weak and fragile; we do not have fangs to fight predators and catch prey or fur to protect us from extreme cold. Culture is the defense mechanism with which evolution equipped us to protect ourselves, so that we can inherit and then pass on our knowledge of manufacturing weapons (to fight predators and catch prey) or clothing and shelter (to protect us from extreme cold). *We don't need fangs or fur, because we have culture.* And just like—despite some minor individual differences—all tigers have more or less the same fangs and all polar bears have more or less the same fur, all human societies have more or less the same culture. Fangs are a universal trait of all tigers; fur is a universal trait of all polar bears; so culture is a universal trait of all human societies. Yes, culture is a cultural universal.

## *Three Examples of Exotic Culture That Never Was*

The recent (and somewhat shameful) history of the social sciences is very instructive in this respect. It shows that every time there was news of a discovery of a new, exotic culture in a remote region of the world, completely different from the Western European culture, it turns out that the discovery was a hoax. Every time, it turns out that there are no human cultures that are radically and completely different from other cultures. We'll share three such examples.

### *Margaret Mead and the Samoa*[16]

In 1923, Margaret Mead (1901–1978), one of the most celebrated anthropologists of all time, was an anthropology graduate student of Franz Boas at Columbia University. Boas was a Jewish refugee from Nazi Germany, and was therefore politically and personally motivated to prove wrong the Nazi policy of eugenics. While this is an admirable goal in and of itself, Boas unfortunately chose the wrong tactics to achieve it. He wanted to show that biology had nothing to do with how humans behave, and that environment—culture—determines human behavior entirely. He was a strong proponent of *cultural determinism.*

In order to demonstrate that culture and socialization determine human behavior in its entirety, Boas gave his graduate students (including Mead) the impossible task of finding a human culture radically different from the Western culture, where people behave completely differently from Americans and Europeans. Margaret Mead was sent to Samoa with this mandate from Boas.

On August 31, 1925, Mead arrived in American Samoa to conduct her research. She was to spend six months doing her fieldwork. Unbeknownst to Boas, however, Mead was involved in another, secret research project, and spent almost all of her time in Samoa doing this other work. She was to leave Samoa in a month, and she had not done any of the fieldwork for Boas on the topic of cultural and behavioral variability to find evidence that Samoan behavior was completely different from American behavior. She decided to finish this work quickly by interviewing two young local women about the sexual behavior of adolescents in Samoa on March 13, 1926.

Mead knew that in the United States and the rest of the Western world, boys were sexually aggressive and actively pursued girls, while girls were sexually coy and waited to be asked out on dates by boys. "How different are things in Samoa? How are Samoan boys and girls when it comes to sex?" Mead asked her two young female informants, Fa'apua'a Fa'amu and Fofoa Poumele.

Fa'apua'a and Fofoa, just like young women everywhere, were quite embarrassed to talk about sex to a total stranger. So they decided to make a big joke about it out of sheer embarrassment. They told Mead the *opposite* of how things were in Samoa. They told her that boys were quite shy, and girls actively pursued boys sexually. It was a hoax, but in the minds of Fa'apua'a and Fofoa, the story that they were telling Mead was so outrageous and so obviously untrue that they couldn't believe anyone in her right mind would believe them.

Except that Mead did, for this was exactly the type of "evidence" that Boas had sent her to Samoa to gather. Here now was evidence that sexual behavior of adolescents could be completely different from (nay, the *opposite* of) how it is in the United States. So culture does completely determine human behavior after all! Mead was ecstatic. She left Samoa in April 1926 and published her "findings" in Samoa in a book called *Coming of Age in Samoa* in 1928. The book immediately became an international bestseller and later a classic in cultural anthropology, and, among other things, formed the foundation of modern feminism. Feminists pointed to the "evidence" in the book to support their claim that, given different "gender socialization," Western boys and girls could be completely different. Boys could be more like girls, and girls could be more like boys. So, in a sense, modern feminism was founded on the basis of a hoax.

More than sixty years later, on May 2, 1988, Fa'apua'a, who was then 86 years old, told a Samoan government official (who happened to be the son of Fofoa, who passed away in 1936) that everything she and her friend Fofoa told Margaret Mead about the sexual behavior of Samoan boys and girls on that fateful night of March 13, 1926, was untrue. It was a hoax. As it turns out, overwhelming ethnographic evidence by now shows that Samoan adolescents are no different from adolescents anywhere else in the world. Boys are sexually aggressive and active, and girls are sexually coy and shy.

## *The Gentle Tasaday*

In 1968, biosocial anthropologist Napoleon Chagnon published the first edition of the anthropology classic *Yanomamö: The Fierce People.*[17] In the book, Chagnon describes the life of a tribe of South American Indians called the Yanomamö, living in the jungles of Brazil and Venezuela. The Yanomamö are so fierce and warlike that a third of adult males (and 7 percent of adult females) die in their constant battle. They are thought to be the fiercest people on earth.

Now that the Yanomamö were known to the world through Chagnon's work, the cultural determinists–the intellectual descendants of Franz Boas–had a task at hand. If human culture and behavior were infinitely variable, then there must exist the opposite of the Yanomamö somewhere on earth. If there were "the fiercest people on earth," then there must also be "the gentlest people on earth." Merely three years later, the cultural determinists got their wish.

In 1971, Manuel Elizalde, an official of the Marcos government in the Philippines, discovered an isolated tribe of twenty-six men, women, and children on the island of Mindanao. Called the Tasaday, they were said to lead a Stone Age life, without any knowledge of agriculture or even the existence of any other humans besides themselves. They had been completely cut off from the rest of the world for centuries. They were wearing leaves and living in a cave. Among other things, they were so peaceful (*so opposite of the Yanomamö*) that their language did not even contain any word for violence, conflict, or aggression. Two years later, a book describing their peaceful life was published with the predictable title *The Gentle Tasaday.*[18]

With the help of the Marcos government, Elizalde tightly controlled media and scientific access to the Tasaday for fifteen years. As a result, not much more was known about them, and what was known about them by the rest of the world was officially sanctioned by Elizalde. In 1986, the Marcos government collapsed and Elizalde fled the country to Costa Rica. When two journalists went to the site of original discovery of the Tasaday, they found the cave empty. They found the Tasaday in a nearby village, wearing T-shirts and blue jeans. Upon further questioning, two of the original twenty-six Tasaday admitted to pretending to be Stone Age people upon Elizalde's insistence. It turns out that Marcos had instructed Elizalde to manufacture this band of peaceful Stone Age people in order to attract the world's attention to the Philippines but away from the brutal policies of his oppressive government. When a group of German journalists went to the cave a few days after the two original journalists uncovered the hoax, they discovered the Tasaday once again playing the parts of Stone Age people, pretending to live in a cave and wearing leaves *on top of their T-shirts and blue jeans.*

When one of us (Kanazawa) took his first sociology course in 1982, his instructor used the second edition, published in 1981, of the bestselling introductory sociology textbook *Sociology* by Ian Robertson. On page 57, there is a picture of the Tasaday, all peacefully and quietly sitting in their cave. The caption to the photograph reads, "The Tasaday, a recently discovered 'stone age' tribe in the Philippines, apparently do not have words in their language to express enmity or hatred. Competition, acquisitiveness, aggression, and greed are all unknown among these gentle people. The existence of societies like the Tasaday challenges Western assumptions about 'human nature.'" Five years later, Kanazawa taught his own introductory sociology course at the University of Washington for the first time and used the third edition of Robertson's still bestselling textbook, published in 1987–a year after the hoax had been uncovered. All references to the Tasaday had been deleted in the third edition.

Incredibly, anthropologists still debate the authenticity of the Tasaday even today,[19] but the majority of opinions appears to be that they were not a genuine Stone Age people. One thing is certain: A small tribe of twenty-six people could not have been completely isolated from the outside world for centuries because that would lead to massive inbreeding. And they also could not possibly have been so peaceful that their language lacked any word for conflict and competition. For better or worse, aggression and violence are part of male human nature. It could be heightened, as among the Yanomamö, but it could not be completely erased from human nature.

### *The Native American Environmentalism*[20]

Unlike the first two, our third and final example of an exotic culture that never was is something that is not yet widely known as false. It is commonly believed even today that, unlike the later European settlers to the American continents, Native Americans are protective of the environment. It is often said that Native Americans make every decision with the next seven generations in mind.

In 1854, the governor of the Washington Territory, on behalf of President Franklin Pierce, met with Chief Seattle, leader of the Duwamish Indians, and offered to buy Chief Seattle's land. This was Chief Seattle's response to the offer:

> How can you buy or sell the sky? The land? The idea is strange to us. . . . Every part of this earth is sacred to my people. Every shining pine needle, every sandy shore, every mist in the dark woods, every meadow, every humming insect. All are holy in the memory and experience of my people. . . . Will you teach your children what we have taught our children? That the earth is our mother? What befalls the earth befalls all the sons of earth. This we know: the earth does not belong to man, man belongs to the earth.

It's a beautiful speech. The only problem is that Chief Seattle never made it. The whole speech was written by a white screenwriter and professor of film, Ted Perry, for the 1971 ABC TV drama *Home.* It was fiction. This is the origin of the myth of Native American respect for the environment.

There is no contemporaneous record of what Chief Seattle actually said at the meeting with the governor in 1854, but according to one eyewitness account, made thirty years later, Chief Seattle thanked the governor for the President's generosity. He was very eager to do business with the President and sell his land to the US government.

The myth that Native Americans are protective of the environment was further fortified by the "Keep America Beautiful" series of public service announcements in 1971, the same year *Home* aired, with the unforgettable image of the "crying Indian." The Indian witnesses white people littering and polluting the environment, and quietly weeps for Mother Earth and the abuse that she must go through at the hands of white people. The message of the public service announcement was that we must all be as protective of the environment as the Native Americans were.

(After his death in 1999, it was revealed that Iron Eyes Cody, the man who played the "crying Indian" in the public service announcements in 1971 and subsequently made a career in Hollywood, portraying numerous Native American characters in movies and TV shows, was not Native American at all. He was born Espera Oscar DeCorti, a son of two *Italian* immigrants.)

Archaeological evidence shows that Native Americans were no more or no less protective of the environment than were any other groups on earth. A large majority of plant and animal species that ever existed on the American continents had been driven extinct by Native Americans long before Columbus set foot in the West Indies. Environmental protection is a luxury that became possible to Western societies only in the last several decades. Before industrialization and the current age of material abundance, all human groups had to exploit the environment to the maximum just to survive. No one could afford to be environmentally conscious, and Native Americans were no exception.

The point of these examples of exotic culture that never was is to highlight the fact that all human cultures, however exotic and seemingly different on the surface, are essentially the same. There are no human cultures that are radically and completely different from any other, just like there are no human bodies that are radically and completely different from any other. Every time there appears to be a new discovery of an exotic culture that is different from all others, it turns out to be a hoax.

## On to the Puzzles and Questions

Now that we have discussed the fundamentals of evolutionary psychology in the last two chapters, you should feel free to delve into the questions that we pose, and answers we suggest for them, in the substantive chapters (chapters 3–8). There is, of course, much more to evolutionary psychology than we discussed in chapters 1 and 2, and if you are interested, we suggest that you explore the books and articles that we recommend in footnote [10] in chapter 1. But our discussion in the last two chapters should be sufficient to inform the questions and answers anywhere in the next six chapters. So feel free to jump in, jump around, and explore the questions that most interest you. Enjoy!

# 3

# Barbie—Manufactured by Mattel, Designed by Evolution

## THE EVOLUTIONARY PSYCHOLOGY OF SEX AND MATING

Because it focuses so much on reproductive success, most of the fascinating studies that come out each year in evolutionary psychology are about sex and mating. One of the earliest studies in the field, conducted in the mid-1980s, surveyed over ten thousand people from thirty-seven cultures throughout the world and asked them what they sought in their ideal mate.[1] To the surprise of everyone (except for evolutionary psychologists), the study found that, regardless of culture, language, religion, race, or geography, men everywhere want the same things in women, and women everywhere want the same things in men (but different from what men want in women).

You may believe that your personal preferences for an ideal mate are truly personal and individual, not shared by other people. The basic message of evolutionary psychology is that, contrary to what you may have thought, your preferences and desires for your ideal mate are strongly shaped by the forces of evolution. Ultimately, it's not what you want that matters; it's what your genes want in order to assist their goal of spreading themselves as much and as far as possible.

Another message of evolutionary psychology, particularly important in this book, is that a lot of the rest of human social behavior–in politics, in religion, in economics–is ultimately about sex. As we try to show in later chapters on these areas of social life, what we commonly think of as political behavior, religious behavior, or economic behavior is essentially about sex and mating. So, in that sense, this chapter is more fundamental than all the other chapters. And, of course, it is the "sexiest" chapter!

## **Q.** Why Do Men Like Blonde Bombshells (and Why Do Women Want to Look Like Them)?

It is commonly believed by those who subscribe to the Standard Social Science Model–in other words, virtually everybody except for evolutionary psychologists–that the media impose arbitrary images of ideal female beauty on girls and women in our society, and force them to aspire to these artificial and arbitrary standards. According to this claim, girls and women want to look like supermodels or actresses or pop idols because they are bombarded with images of these women. By implication, according to this view, girls and women will cease to want to look like them if the media would cease inundating them with such images, or else change the arbitrary standards of female beauty.

*Nothing could be further from the truth.* To claim that girls and women want to look like blonde bombshells because of the billboards, movies, TV shows, music videos, and magazine advertisements makes as little sense as to claim that people become hungry because they are bombarded with images of food in the media. If only the media would stop inundating people with images of food, they would never be hungry! Anyone can see the absurdity of this argument. We become hungry periodically because we have physiological and psychological mechanisms that compel us to seek and consume food. And we have these innate mechanisms because they solve an important adaptive problem of survival. Our ancestors (long before they were humans or even mammals) who somehow did not become hungry for food did not survive long enough to leave offspring who carried their genes. We would of course become hungry just as much even if all the commercials about food disappeared today. The advertisements are the *consequences* of our tendency to become hungry, not the causes. They exploit our innate needs but do not create them.

The same is true with the ideal of female beauty. Two pieces of evidence should suffice to refute the claim that images in the media, and "culture" in general, force girls and women to desire to look like blonde bombshells. First, as we note below, women were dying their hair blonde more than half a millennium, possibly two millennia, ago, when there were no TV, movies, and magazines (although there were portraits, and it is due to these portraits that we know today that women were dying their hair blonde in fifteenth- and sixteenth-century Italy).[2] Women's desire to be blonde preceded the media by centuries, if not millennia.

Second, a recent study shows that women in Iran, where they are generally not exposed to the Western media and culture, and thus would not know Jessica Simpson from Roseanne Barr, and where most women wear the traditional Muslim *hijab* that loosely covers their entire body so as to make it impossible to tell what shape it is, are actually *more* concerned with their body image and want to lose *more* weight than their American counterparts in the land of *Vogue* and the Barbie doll.[3] The Standard Social Science Model, which ascribes the preferences and desires of women entirely to socialization by the media, would have difficulty explaining how Italian women in the fifteenth century and Iranian women today both aspire to and pursue the same ideal image of female beauty as do women in contemporary Western societies.

Why, then, do women want to look like blonde bombshells? Evolutionary psychology suggests that it is because men want to mate with women who look like them. Women's desire to look like them is a direct, realistic, and sensible response to this desire of men. This then begs the question: Why do men want to mate with women who look like them? Because women who look like them have higher reproductive value and fertility and attain greater reproductive success on average. There is nothing arbitrary about the image of ideal female beauty; it has been precisely and carefully calculated by millions of years of evolution by sexual selection. Men today want to mate with women who look like blonde bombshells, and as a result, women want to look like them, because our ancestral men who did not want to mate with women who looked like them did not leave as many offspring.

Let's take a closer look at exactly what we mean by "blonde bombshells." Note, first, that there has been a long line of blonde bombshells in the Western media: Pamela Anderson, Madonna, Brigitte Bardot, the popular British bombshell Jordan–all the way back to the iconic Marilyn Monroe and even further back in history. And there are contemporary examples as well: Jessica Simpson, Cameron Diaz, Scarlett Johansson, among many others. Readers from non-Western societies can suitably substitute representatives of female beauty from their own cultures. We do not know who they are, but we can nonetheless be confident that they share many features with their Western counterparts.

What are these features? We will isolate and discuss in turn the key features that define the image of ideal female beauty. These are youth, long hair, small waist, large breasts, blonde hair, and blue eyes. There is evolutionary logic behind each one.

### *Youth*

Men prefer young women because they have greater reproductive value and fertility than older women. A woman's *reproductive value* is the expected number of children that she will have in the remainder of her reproductive career, and therefore reaches its maximum at the onset of menstruation, steadily declines over her life course, and reaches zero at menopause.[4] Her *fertility* is the average number of children that she actually has at any given age, and reaches its maximum in her twenties. Evolutionary psychological logic suggests that this is why men are attracted to teenage girls and young women, despite the laws of civilized society concerning the age of consent. Remember, there were no laws against statutory rape in the ancestral environment; in fact, there were no laws at all. The Savanna Principle, which states that the human brain has difficulty dealing with entities that did not exist in the ancestral environment, suggests that the human brain cannot really comprehend written laws, including laws regarding the age of consent.

For example, male high school teachers and college professors in the United States (but *not* their female colleagues) have a higher-than-expected rate of divorce and a lower-than-expected rate of remarriage, probably because they are constantly exposed to girls and women at the peak of their reproductive value. Any adult woman they might be married to or date pales in comparison to their female students on the reproductive score.[5] This can also explain why most Hollywood marriages do not last very long. Actors are constantly exposed to and closely associate with younger and younger generations of starlets, while their actress-model wives can only get older.

### *In (Futile) Search for the Human Barbie Doll*

Here's a little autobiographical aside, which nonetheless makes our point about the importance of youth in the ideal female beauty. When we first began writing this book in 2000, we chose Pamela Anderson as the ideal of female beauty, the human Barbie doll, and the title of this section was "Why Do Men Like Pamela Anderson (and Why Do Women Want to Look Like Her)?" As years went by, however, she ceased to fit the bill. *Baywatch* went off the air in 2001, and Pamela Anderson turned 40 in 2007. So we then chose to replace her with Britney Spears, who was at the time the perfect image of a virginal, nubile princess. Well, you know what has happened to her lately. Next candidate, please!

As we sought yet again to replace Britney Spears with another perfect image of female beauty, it dawned on us that, no matter whom we would choose to use, she would be out of date pretty soon because of the high premium placed on youth for the ideal female beauty. (Had we written this book thirty years ago, this section would have been titled "Why Do Men Like Farrah Fawcett-Majors [and Why Do Women Want to Look Like Her]?" It would have made our book look *really* dated by now; Farrah Fawcett turned 60 in 2007.) Since we want our book to be read for a long time and don't ever want it to look dated, we finally decided not to use an actual example of a blonde bombshell.

### *Long Hair*

Men in general prefer women with long hair.[6] And most young women choose to grow their hair long. Once again, men's preference for women with long hair is probably the reason for women's preference to grow their hair long. The question thus is: Why do men prefer women with long hair?

Because the human fetus grows inside the woman's body for nine months, and then the mother nurses the newborn baby for a few years afterward, the woman's health is crucial for the well-being of the child. Sickly women do not make good mothers, to a significantly greater extent than sickly men do not make good fathers. Thus, men are interested in selecting healthy women to be the mothers of their children. Part of the reason that men prefer young women, besides their higher reproductive value and fertility, is that younger women tend to be healthier on average than older women.

How can men assess the health of their potential mates? There were no clinics in the ancestral environment; ancestral men had to judge women's health by themselves. One accurate indicator of health is physical attractiveness, and this is the reason why men like beautiful women. (See the section, "Why Is Beauty *Not* in the Eye of the Beholder or Skin-Deep?" later in this chapter.) Another good indicator of health is hair. Healthy people (men and women) have lustrous, shiny hair, whereas the hair of sickly people loses its luster. During illness, a body needs to sequester all available nutrients (like iron and protein) to fight the illness. Since hair is not essential to survival (compared to, say, bone marrow), hair is the first place to which a body turns to collect the necessary nutrients. Thus, a person's poor health first shows up in the condition of the hair.[7]

Further, hair grows very slowly, at about six inches per year. That means that if a woman has shoulder-length hair (two feet long), it accurately indicates her health status for the past four years, because once the hair grows there is nothing the bearer can do to change its appearance later. A woman might be healthy now, but if she was sick sometime in the past four years, her long hair would indicate her past sickly status. And there was nothing a woman could do in the ancestral environment to make her hair appear healthy and lustrous when she was not healthy. This is also why older women tend to keep their hair short, because they tend to become less healthy as they grow older, and they do not want telltale signs of their current health status hanging from their heads.

If you want to see this process in action, try a little experiment on your own. Find a female stranger in a public place (like a park or a subway station). Observe her from behind, without looking at her face, her hands, her clothes, or anything else about her, and look only at her hair. Try to guess her age from the condition of her hair alone, nothing else. Once you come up with a guess for her age, pass by her, turn around to the front, and discreetly look at the woman's face. You will find that you are very rarely surprised by her apparent age when you look at her face and her entire body, because the condition of her hair is usually a very accurate indicator of her age. You've now discovered the importance of hair as an indicator of age in the ancestral environment.

### *Small Waist*

Why are 36–24–36 considered the ideal female measurements? It turns out that these numbers are not chosen arbitrarily.

An evolutionary psychologist at the University of Texas, Devendra Singh, conducted experiments in different societies to demonstrate that men have a universal preference for low waist-to-hip ratio (the waist measurement divided by the hip measurement). Presented with figure drawings of women identical in every way except the waist-to-hip ratio (varying from 0.7 to 1.0), most men in Singh's experiments expressed preference for women with the waist-to-hip ratio of 0.7, which is very close to the waist-to-hip ratio of anyone with the 36–24–36 measurements (0.67).[8] One of us (Kanazawa) has informally replicated Singh's experiments in three different countries on three different continents (the US, New Zealand, and the UK) and found the same results as Singh. Most men prefer women with a 0.7 waist-to-hip ratio, and most women prefer men with a 0.9 waist-to-hip ratio.

Why, then, do men want women with low waist-to-hip ratios? Singh argues that this is because healthy women have lower waist-to-hip ratios than unhealthy women. A host of diseases–such as diabetes, hypertension, heart attack, stroke, and gallbladder disorders–change the distribution of body fat so that sickly women cannot maintain low waist-to-hip ratios. Women with low waist-to-hip ratios are also more fertile; they have an easier time conceiving a child and do so at earlier ages because they have larger amounts of essential reproductive hormones.[9] And, of course, women who are already pregnant with another man's child cannot maintain a low waist-to-hip ratio.

The female waist-to-hip ratio also fluctuates, albeit very slightly, over the menstrual cycle; it becomes lowest during ovulation, when the woman is fertile.[10] Thus, men are unconsciously seeking healthier and more fertile women when they seek women with small waists.

The preference for a low waist-to-hip ratio, identified by Singh, explains both the popularity of corsets in many Western societies throughout history as a device to make women's waists appear as small as possible, and the current trend of young women to bare their midriffs. It also explains why it is teenage girls, not menopausal women, who are more likely to bare their midriffs as an honest signal of their high fecundity (the ability to conceive), just like it is young women, not old women, who grow their hair long as an honest signal of their health. Once again, the superstardom of Britney Spears was not the *cause* of young girls' desire to show their midriffs; rather, it is a consequence of it.

### Large Breasts

Why men prefer women with large breasts had long been a mystery in evolutionary psychology, especially since the size of a woman's breasts has no relationship with her ability to lactate; women with small breasts can produce as much milk for their infants as those with large breasts.[11] So women with large breasts do not necessarily make better mothers than women with small breasts. Why, then, do men prefer women with large breasts? There was no satisfactory answer to this question until recently.

The then Harvard anthropologist Frank Marlowe suggested a solution to this puzzle in the late 1990s,[12] although with hindsight it is another mystery why nobody else thought of the idea sooner. Marlowe makes the simple observation that larger, and hence heavier, breasts *sag* more conspicuously with age than do smaller breasts. Thus, it is much easier for men to judge a woman's age (and her reproductive value) by sight if she has larger breasts than if she has smaller breasts, which do not change as much with age. Recall that there were no driver's licenses or birth certificates that men could check to learn how old women were in the ancestral environment. There was no calendar and thus no concept of birthdays in the ancestral environment, so women themselves didn't know exactly how old they were. The ancestral men needed to infer a woman's age and reproductive value from some physical signs, and the state of her breasts provided a pretty good clue, but only if they were large enough to change their shape conspicuously with age. Men could tell women's ages more accurately, and attempt to mate with only young women, if they had larger breasts. Marlowe hypothesizes that this is why men find women with large breasts more attractive.

More recently, there has been a competing evolutionary psychological explanation for why men prefer women with large breasts. A study of Polish women shows that women who simultaneously have large breasts and a tight waist have the greatest fecundity, indicated by their levels of two reproductive hormones (17-ß-estradiol and progesterone).[13] So men may prefer women with large breasts for the same reason as they prefer women with small waists. Further empirical evidence is necessary to evaluate which of these two competing evolutionary psychological explanations is more accurate. This is just one of many areas where there are competing hypotheses in evolutionary psychology–a sign of active, healthy science.

Men can accurately infer a woman's age and reproductive value if they can directly observe their breasts and other physical features (such as the fat content of the body). But what would men do if they could not directly observe women's bodies? What if the woman's body is concealed, by heavy clothing, for instance? Men need another way to determine a woman's age: her hair color.

## *Blonde Hair*

Why do blondes have more fun? Because gentlemen prefer blondes.[14] Why do gentlemen prefer blondes? Because they have evolved psychological mechanisms that predispose them to prefer women with blonde hair. Why?

The notion that blonde hair is the female ideal goes back at least half a millennium,[15] possibly a couple of millennia.[16] There is evidence that women during the Roman era and the Renaissance period dyed their hair blonde, long before the discovery of peroxide in 1812. Women desired to be blonde so strongly throughout recorded history that they accomplished it without the aid of peroxide.

Some believe that men prefer blonde hair because blonde women tend to have lighter skin, which they prefer.[17] But this seems to be false. While men do prefer women with lighter skin color,[18] because it is an indication of higher fertility[19] (a woman's skin color darkens when she is pregnant or on the Pill),[20] the lightest skin color is associated with red hair, not blonde hair; yet, according to one study,[21] both men and women have *extreme* aversion to potential mates with red hair. It turns out that men prefer blonde hair for exactly the same reason that they prefer large breasts: both are accurate indicators of a woman's age and thus reproductive value.

What distinguishes blonde hair from all other hair colors is that it changes dramatically with age. Young girls with light blonde hair usually grow up to be women with brown hair (although there are a very few women who retain their light blonde hair into adulthood). Thus, if men prefer to mate with women with blonde hair, they are unconsciously attempting to mate with younger (and hence, on average, healthier and more fecund) women with greater reproductive value and fertility. It is no coincidence that blonde hair evolved in Scandinavia and northern Europe, where it is very cold in winter. In Africa, where our ancestors evolved for most of their evolutionary history, people (men and women) mostly stayed naked. In such an environment, men could accurately assess a woman's age by the distribution of fat on her body or by the firmness of her breasts (as discussed above). Men in cold climates did not have this option, because women (and men) bundled up in such environments. This is probably why blonde hair evolved in cold climates as an alternative means for women to advertise their youth.[22] Men then evolved a predisposition to prefer to mate with women with blonde hair in response, because those who did on average had greater reproductive success than those who did not, because, unbeknownst to them, they ended up mating with younger, healthier women with greater reproductive value and fertility.

Incidentally, this also suggests that the stereotype that blondes are dumb may have some statistical basis. Why do people believe that blondes are dumb? Recall that the human brain, including the stereotypes that it generates, is adapted to the ancestral environment. What would be the average age of light blondes in the ancestral environment (say, northern Europe ten thousand years ago) in the absence of hair dye? Roughly 15. What would be the average age of brunettes in the same environment? Roughly 35. A 15-year-old woman is bound to be more naïve and less experienced, mature, and wise (in other words, "dumb") than a 35-year-old woman, no matter what her hair color. It's not that blonde women are dumber than brunette women; it's that younger women are "dumber" (less knowledgeable) than older women, and blonde hair is a reliable indicator of extreme youth. The same logic is probably behind the stereotype that women with large breasts are dumb. In the ancestral environment, without plastic surgery or even bras, only very young women had large, firm breasts.

### *Blue Eyes*

The typical description of ideal beauty always goes "blonde, blue eyes." After Marlowe proposed a solution for the mystery of why men prefer women with large breasts, the attraction to blue eyes remained the only mystery to solve in the area of traits associated with physical attractiveness. We now knew why men preferred women with all the traits that characterize Barbie or a typical blonde bombshell, and we knew the evolutionary logic behind each of them. But eye color, more so than hair color, seems such an arbitrary trait. Why should women with blue eyes be any different from those with green or brown eyes? Yet a preference for blue eyes seems both universal and undeniable.[23]

There is an additional layer to the mystery of blue eyes. Unlike all the other traits discussed above, which are considered to be attractive *only* for women, blue eyes are thought to be attractive for both women *and* men. For instance, a typical description of an attractive man is "tall, *dark*, and handsome," not blond; unlike blonde women, blond men are not universally considered to be attractive (because women generally prefer to mate with older, not younger, men).[24] Yet, as the examples of Frank Sinatra ("Ol' Blue Eyes") and Paul Newman show, men with blue eyes are considered to be attractive, just like women with blue eyes. So it appears that the answer to the question, "Why are blue eyes attractive?" must involve more than male sexual preference.

The attraction of blue eyes remained an evolutionary mystery until an undergraduate student, Lee Anne Turney, suggested a novel solution in her term paper for a class that she took from one of us (Kanazawa) in spring 2002. As far as we know, hers is the only available explanation for the attraction of blue eyes that anyone has ever proposed, and it has at least surface plausibility. But, of course, it will have to be subjected to rigorous experimental testing before it can become an accepted explanation.

Turney points out that the human pupil dilates when an individual is exposed to something that he or she likes.[25] For instance, the pupils of women and infants (but not men) spontaneously dilate when they see babies. Thus, pupil dilation, which is usually beyond an individual's conscious volitional control, can be used as an honest indicator of interest and attraction. Most people are not even aware that the size of their pupils changes when they see something they like, so it would be difficult to deceive others by consciously manipulating their pupil size. We cannot help but reveal our interest in and attraction to others through the size of our pupils.

Turney then makes two simple observations. First, every human pupil is dark brown, regardless of the color of the iris, which encloses the pupil and determines the color of the eye. Second, blue is the *lightest* color of human iris. The consequence of these two observations is that *the size of the pupil is easiest to determine in blue eyes.* If you face people with different eye colors and must determine whether each person likes or is interested in you, with all else equal, it is easiest to read the blue-eyed person's level of interest or attraction. Turney's argument, which we believe might be true, is that blue-eyed people are considered attractive as potential mates because it is easiest to determine whether they are interested in us or not. It is easier to "read the minds" of people with blue eyes than of those with eyes of any other color, at least when it comes to interest or attraction.

One of the advantages of Turney's solution is that it not only explains why blue eyes are considered to be ideal in mates but also explains why, unlike all the other traits we discuss in this section, blue eyes are considered to be attractive in both sexes. It is just as important for women to read men's minds as it is for men to read women's; if anything, given the far greater consequences of making mistakes by being attracted to the "wrong" person, women should have a greater need to decide whether their potential mate's seeming interest in them is genuine or not. The negative consequences of being fooled by a lying suitor are much greater for women, so blue eyes should be a more important characteristic in men than in women.

Incidentally, we believe that Turney's logic can also explain why people with dark brown eyes are often considered to be "mysterious." They are mysterious because their minds–that is, whether or not they are interested in or attracted to us–are more difficult to assess. The color of the dark brown iris is very similar to the (universal) color of the pupil, and so it is very difficult to gauge the size of the pupil in dark brown eyes. Many people, both men and women, express dislike for extremely dark brown eyes.[26]

## *The Big Irony: Why Modern Men Are Fooled*

So men like women who look like blonde bombshells, or Barbie, and women want to look like them, because each of their key features (youth, long hair, small waist, large breasts, blonde hair, and blue eyes) is an indicator of youth and thus of health, reproductive value, and fertility. There is precise evolutionary logic behind the image of ideal female beauty. By now, astute readers may have caught on to the irony of it all. *None of what we have said above is true any longer.* Through face-lifts, wigs, liposuction, surgical breast augmentation, hair dye, and color contact lenses, any woman–regardless of age–can have many of the key features that define the ideal female beauty. Very little of Pamela Anderson's appearance is natural. A 40-year-old woman today can rely on modern technology to continue to look like a 20-year-old woman. Farrah Fawcett at 60 looks better than most "normal" women half her age.

And men fall for them. As the Savanna Principle suggests, their brains cannot really comprehend silicone breasts or blonde hair dye, because these things did not exist in the ancestral environment ten thousand years ago. Men can cognitively understand that many blonde women with firm large breasts are not actually 15 years old, but they still find them attractive because their evolved psychological mechanisms are fooled by the modern inventions that did not exist in the ancestral environment.

## Q. Why Is Beauty *Not* in the Eye of the Beholder or Skin-Deep?

They say beauty is in the eye of the beholder, which means that different people possess different standards of beauty and that not everyone agrees on who is beautiful and who is not. They also say beauty is skin-deep, which means that there are no real differences between attractive and unattractive people besides their looks. Both of these sayings make perfect sense from the perspective of the Standard Social Science Model. Since humans are born with blank slates for minds, the logic goes, everything, including tastes for and standards of beauty, must be acquired after birth through socialization. As different people have different life experiences in different cultures, they naturally acquire different standards of beauty. Some features are considered to be beautiful by some people in some cultures, and completely different features are considered beautiful by other people in other cultures. The differences between attractive and unattractive people are therefore arbitrary.

On the surface, both the sayings "beauty is in the eye of the beholder" and "beauty is only skin-deep" and the Standard Social Science Model explanations for them appear plausible. Many introductory sociology and anthropology textbooks, for example, include pictures of people who are considered to be beautiful in different cultures, and some of them look quite bizarre to the contemporary Western eye. However, evolutionary psychological research has once again overturned this common assumption and widespread belief.

As it turns out, the standards of beauty are universal, both across individuals in a single culture and across all cultures.[27] Within the United States, both East Asian and white individuals[28] and white and black individuals[29] agree on which faces are more or less attractive. Cross-culturally, there is considerable agreement in the judgment of beauty among East Asians, Hispanics, and Americans;[30] Brazilians, Americans, Russians, the Aché of Paraguay, and the Hiwi of Venezuela;[31] Cruzans and Americans in Saint Croix;[32] white South Africans and Americans;[33] and the Chinese, Indians, and the English.[34] In none of these studies does the degree of exposure to the Western media have any influence on people's perception of beauty. If, as the Standard Social Science Model contends, the standards of beauty are acquired and learned through socialization within different families and cultures, how is it possible for people from such diverse cultures to agree broadly on who is beautiful and who is not?

It appears that people from different cultures share the same standards of beauty because they are *innate*. Two studies conducted in the mid-1980s independently demonstrate that infants as young as two and three months old gaze longer at a face that adults judge to be more attractive than at a face that adults judge to be unattractive, indicating the infants' preference for attractive faces.[35] In the most recent version of this experiment, newborn babies *less than one week old* show significantly greater preference for faces that adults judge to be attractive.[36] Another study shows that 12-month-old infants exhibit more observable pleasure, more play involvement, less distress, and less withdrawal when interacting with strangers wearing attractive masks than when interacting with strangers wearing unattractive masks.[37] They also play significantly longer with facially attractive dolls than with facially unattractive dolls. The findings of these studies are consistent with the personal experiences and observations of many parents of small children, who find that their children are much better behaved when their babysitters are physically attractive than when they are not.

Because even the most ardent proponents of the Standard Social Science Model would admit that one week (or even a few months) is not nearly enough time for infants to have learned and internalized the cultural standards of beauty through socialization and media exposure, these studies strongly suggest that the broad standards of beauty might be innate, not learned or acquired through socialization. The balance of evidence seems to indicate that beauty is decidedly *not* in the eye of the beholder, but might instead be part of universal human nature. In other words, "beauty is in the adaptations of the beholder."[38]

But this simply begs the question: Why are the standards of beauty innate? Why are we born with a common general perception as to who is beautiful and who is not? What is the evolutionary logic behind this?

There appear to be a few features that characterize physically attractive faces: bilateral symmetry, averageness, and secondary sexual characteristics.[39] Attractive faces are more symmetrical than unattractive faces.[40] Bilateral symmetry (the extent to which the facial features on the left and the right sides are identical) decreases with exposure to parasites, pathogens, and toxins during development,[41] and with genetic disruptions such as mutations and inbreeding.[42] Developmentally and genetically healthy individuals have greater symmetry in their facial and bodily features, and are thus more attractive. For this reason, across societies, there is a positive correlation between parasite and pathogen prevalence in the environment and the importance placed on physical attractiveness in mate selection; people place more importance on physical attractiveness when there are more pathogens and parasites in their local environment.[43] This is because in societies where there are a lot of pathogens and parasites in the environment, it is especially important to avoid individuals who have been afflicted with them when selecting mates.

Facial averageness is another feature that increases physical attractiveness; faces with features closer to the population average are more attractive than those with extreme features.[44] In the memorable words of Judith H. Langlois, who originally discovered that the standards of beauty might be innate, "attractive faces are only average."[45] Evolutionary psychological reasons for why average faces in the population are more attractive than extreme faces are not as clear as the reasons for why facial symmetry is attractive. Current speculation is that facial averageness results from the heterogeneity rather than the homogeneity of genes. Individuals who have two different copies (or *alleles*) of a gene are more resistant to a larger number of parasites, less likely to have two copies of deleterious genes, and at the same time more likely to have statistically more average faces with less extreme features.[46] If this speculation is correct, it means that, just like bilateral symmetry, facial averageness is an indicator of genetic health and parasite resistance.

Far from being merely in the eye of the beholder or skin-deep, beauty appears to be an indicator of genetic and developmental health, and therefore of mate quality; beauty is a "health certification."[47] More attractive people are healthier,[48] have greater physical fitness,[49] live longer,[50] and have fewer lower back pain problems[51] (although some dispute this conclusion).[52] Bilateral symmetry measures beauty so accurately that there is now a computer program that can calculate someone's level of symmetry from a scanned photograph of a face (by measuring the sizes of and distances between various facial parts) and assign a single score for physical attractiveness, which correlates highly with scores assigned by human judges.[53] A computer program can also digitally average human faces.[54]

Beauty, therefore, appears to be an *objective* and *quantitative* attribute of individuals, like height and weight–both of which were also more or less "in the eye of the beholder" before the invention of the yardstick and the scale.

## Q. Why Is Prostitution the World's Oldest Profession, and Why Is Pornography a Billion-Dollar Industry?

It has often been remarked that female prostitutes cater to men, while male prostitutes also cater to men. Gay or straight, virtually all clients of prostitution are men, and very few women contract the service of prostitutes. Why the sex difference? And why is prostitution the "world's oldest profession"?

Because of the asymmetry in reproductive biology, men's reproductive success is primarily constrained by the number of women to whom they have sexual access, whereas women's reproductive success does not increase linearly with the number of men to whom they have sexual access.[55] For example, if a man has sex with one thousand women in a year, he can potentially produce one thousand children, and realistically about thirty in a year. (The probability of conception per intercourse is about 0.03.)[56] In sharp contrast, if a woman has sex with one thousand men in a year, she can have only one child (barring a multiple birth) in the same time period, which she can achieve by having regular sex with only one man. The probability of a conception if a woman has sex with one man one hundred times (twice a week for a year) is 0.95.[57] So, unlike for men, there is very little reproductive benefit for women in seeking a large number of sex partners.

For this reason, men are selected to desire a far larger number of sex partners than women do.[58] On average, young men profess to desire about eight different sex partners in two years, whereas young women profess to desire only about one in the same time period. This is why men desire *sexual variety* (sexual intercourse with a large number of partners) to a far greater extent than women do.

Prostitution, the world's oldest profession, is simply a consequence of men's evolved desire for sexual variety. It could not have survived millennia as an industry if men did not have such a desire, and women's lack of the same desire—at least to the same extent—explains why prostitution catering to women has not emerged.

Another indicator of sex differences in the desire for sexual variety is in sexual fantasies, which are expressions of both male and female sexual desires entirely unconstrained by reality.[59] Thirty-two percent of young men surveyed in one study report that they have fantasized about sexual encounters with more than one thousand partners in their lifetime, whereas only 8 percent of women report the same variety of partners in their sexual fantasies. Further, men are far more likely than women to switch one imagined partner for another during a single sexual fantasy, thus allowing them to have (albeit imagined) sexual access to multiple partners. As for women's fantasies, given their relatively greater tendency to desire sex in committed relationships rather than in anonymous encounters,[60] women tend to prefer romance novels, not pornography, as their means of fulfilling sexual fantasy.[61]

### *Speaking of Fantasy . . . What about Pornography?*

As with prostitution, an overwhelming majority of consumers of pornography worldwide are men.[62] Given their greater desire for sexual variety, it is understandable why men would consume more pornography and seek out sexual encounters with numerous women in pornographic photographs and videos, as they do when they contract prostitutes in search of greater sexual variety. Unlike consorting with prostitutes, however, watching pornography does not lead to actual sexual intercourse, but the Savanna Principle suggests that a man's brain does not really know that. When men see images of naked and sexually receptive women in photographs and videos, their brains cannot truly comprehend that they are artificial images of women that they will likely never meet, much less have sex with, because no such images existed in the ancestral environment; every single naked and sexually receptive woman that our male ancestors saw was a potential sex partner. As a result, their brains think that they might have actual sexual encounters with these women. Why else would men have an erection when they view pornographic photographs and videos, when the only biological function of an erection is to allow men to have intercourse with women? If men's brains truly comprehended that they would likely never have sex with the naked women in pornography, they would not get an erection when they watch it.

The Savanna Principle applies equally to women as it does to men; women's brains have the same limitations as those of men. This is why women do not consume pornography nearly as much as men do, even though women enjoy having sexual fantasies as much as men do.[63] Women do not seek sexual variety because their reproductive success does not increase by having sex with a large number of partners. In fact, given the limited number of children they can have in their lifetimes, the potential cost of having sex with the wrong partner is far greater for women than it is for men. This is why women are far more cautious about having sex with someone they do not know well; women tend to require a much longer period of acquaintance before agreeing to have sex than men do. The average woman would begin to consider having sex with someone only after she had known him for six months; for the average man, it only takes one week.[64]

So it makes perfect sense for women to avoid casual sex with anonymous strangers, and their brains cannot really tell that there is no chance that they might copulate with a large number of the naked and sexually aroused men they see in pornography. Women's brains do not fully comprehend that they will not get pregnant by watching pornography, just as men's brains do not know that they cannot copulate with women in pornography. Women avoid pornography for the same reason that men consume it; in both cases, their brains cannot really distinguish between real sex partners and the imaginary ones.

## Q. Why Sean Connery and Catherine Zeta-Jones, but Not Lauren Bacall and Brad Pitt?

When the movie *Entrapment* was released in 1999, it caused an uproar among feminists because of the large age difference between the two main characters, who get romantically involved. Sean Connery was 69; Catherine Zeta-Jones was 30. *Entrapment* is hardly the only movie that incurs feminist wrath for the same reason. In the 1993 movie *In the Line of Fire*, Clint Eastwood is 63 while Rene Russo is 39. Feminists charge that these movies reinforce the cultural norm that women have to be young to be desirable, whereas men can be much older and still attractive to women.

To the feminists' chagrin, however, the pattern of an older man and a younger woman in romance is not limited to movies that appeal mostly to men. The pattern is the same in the so-called chick flicks popular among mostly female audiences. For example, in the 1998 movie *Six Days Seven Nights*, Harrison Ford is 56 while Anne

Heche is 29. In *The Horse Whisperer* of the same year, Robert Redford is 61 while Kristin Scott Thomas is 38. In the 1997 blockbuster *As Good as It Gets,* Jack Nicholson is 60 whereas Helen Hunt is 34. This movie was nominated for the Academy Award for Best Picture, and both Nicholson and Hunt won Oscars for their leading roles. Nor is this pattern of older man and younger woman a recent Hollywood trend. In the 1963 classic *Charade*, Cary Grant is 59 while Audrey Hepburn, who actively pursues him romantically, is 34. In *The Big Sleep* of 1946, Humphrey Bogart is 47, and Lauren Bacall is 22. Of course, Bogart and Bacall were married to each other in real life.

With the notable exception of *The Graduate*, it appears that the man is older, sometimes by decades, than the woman among couples in movies. Why is this? Why do both men and women in every generation expect and want the leading male character in movies to be so much older than his female counterpart? The conventional social science explanation relies on cultural norms and socialization. Our "culture" imposes arbitrary standards of desirability, which include being young in the case of women but not men. People in our "culture" are therefore socialized to expect attractive women in movies to be young, not old, whereas men, who are not subject to the same arbitrary standards, can be old and still sexy.

As we discussed earlier in this chapter (see "Why Do Men Like Blonde Bombshells [and Why Do Women Want to Look Like Them]?" above), there is evolutionary logic behind every aspect of ideal female beauty, including youth. With respect specifically to movies, however, there are two pieces of evidence that contradict the conventional view.

First, even though they are produced in the United States, Hollywood movies are now exported throughout the world. And blockbusters in the United States almost always become commercial successes in other countries where they are shown.[65] While repressive nations like China and those in the Arab world might censor the content of Hollywood movies for sexual explicitness and other taboos like homosexuality, there has not been a single case where such regimes censored or banned movies because of the large age difference between the male and the female leads. The premise of large age differences appears to be readily accepted throughout the world.

Second, while movie production throughout the world is heavily dominated by Hollywood and therefore the United States, all cultures produce literature, which often becomes the basis of movies. And it turns out that literary themes and plots in all cultures and throughout recorded history are remarkably similar.[66] While we have not seen data on movies produced outside the United States, we are very confident in predicting that such movies, like the growing number of "Bollywood" movies out of India, also mostly depict romantic scenarios in which the man is considerably older than the woman, and that very few movies (or novels, for that matter) produced anywhere in the world would have the opposite type of couple as romantic leads.

If not cultural socialization, what then accounts for the popularity of romantic couples where the man is much older than the woman? From the evolutionary psychological perspective, it is a direct consequence and reflection of evolved male and female natures. Data collected from societies throughout the world show that men in every single culture prefer to mate with younger women, and *women prefer to mate with older men.*[67] Men prefer younger women because they have greater reproductive value and fertility than older women, and women prefer older men because they possess greater resources and higher status than younger men in every human society.

Further, the older men get, the greater the age difference between them and their desired mates. Men in their twenties want women who are about five years younger than them, whereas men in their fifties want women who are about fifteen years younger. In yet another example of the "exception that proves the rule," the only category of men who prefer to mate with older women are teenage boys.[68] For them, older, not younger, women have greater fertility. In other words, regardless of *their* age, men always prefer to mate with women in their twenties, at the peak of fertility. Women do not show the same pattern; regardless of their age, women prefer men who are about ten years older than them.[69] Since movie producers and authors are in the business of making money by producing stories that appeal to moviegoers and readers alike, it is natural that their products reflect the evolved desires of their target audiences.

It is interesting to note as an aside that while Mrs. Robinson in *The Graduate* was supposed to be much older than Benjamin, being the mother of his girlfriend, Elaine; in reality, Anne Bancroft is only six years older than Dustin Hoffman. It appears that we expect the woman to be so much younger than the man in movies that when the woman is actually older than the man, even by only six years, she is considered to be "too old" for him. We should also note that, by the end of the movie, Benjamin (the Dustin Hoffman character in *The Graduate*) ends up with the young Elaine, not with her mother, consistent with the evolutionary psychological prediction. Sorry for the spoiler.

## **Q.** He Said, She Said: Why Do Men and Women Perceive the Same Situation Differently?

Miscommunication and misunderstanding between men and women, created when a man and a woman perceive the same situation entirely differently, are the staple of television situation comedies and literary novels alike. A man and a woman encounter each other and have a pleasant and friendly conversation. The man is thinking that she is romantically and sexually attracted to him, while the woman is entertaining no such thought; she is simply being nice. Whether you're male or female, you can probably think of a real situation in your own life that involved this type of miscommunication with someone of the opposite sex.

The phenomenon is not merely anecdotal or only in our personal experiences; it has been scientifically documented.[70] In a laboratory experiment, a male and a female participant engage in a five-minute conversation, while, unbeknownst to both participants, a male and a female observer watch the entire interaction. After the interaction, both the male participant and the male observer rate the female participant as being more promiscuous and seductive than do the female participant or the female observer. Male participants report being more sexually attracted to their conversation partner than do female participants. So men and women can engage in or observe the same situation, but men perceive greater likelihood of sexual relationship than women do.

The failure to recognize this pervasive sex difference in cognitive biases can lead to costly mistakes, not only for individuals but for corporations and society alike. In January 1998, the American supermarket chain Safeway (not related to the British supermarket chain of the same name, which has recently been acquired by the rival chain Morrisons) started implementing the "superior customer service policy," which required all Safeway employees to look customers in the eye and smile.[71] If the customers paid by check or credit card, cashiers were required to quickly scan the customer's last name and thank them personally, as in "Thank you, Mr. So-and-so, for shopping at Safeway" while looking them in the eye and smiling.

Reflecting the unquestioned assumptions of the Standard Social Science Model, the Safeway company policy was completely egalitarian. It required both male and female employees to greet both male and female customers in the identical "friendly" manner. And the policy worked very well roughly three-quarters of the time, between a male employee and a male customer, between a male employee and a female customer, and between a female employee and a female customer. However, the policy backfired when the employee was female and the customer was male. When the female employee gazed deeply into the customer's eyes, smiled, and thanked him by name, some male customers "naturally" assumed that she was attracted to him and started harassing her by following her around–both at and away from work. Eventually, five female employees had to file a federal sex discrimination charge against Safeway to force it to stop this policy, which the corporation did when it reached an out-of-court settlement.

Why do these problems happen? Why are men more likely to infer sexual interest in a neutral encounter than women are? Two evolutionary psychologists, Martie G. Haselton and David M. Buss, offer an explanation in their Error Management Theory.[72]

### *The Cost of Misreading the Signals*

This theory begins with an observation, made earlier by others,[73] that decision making under uncertainty often results in erroneous inferences, but some errors are more costly in their consequences than others. Natural and sexual selection should then favor the evolution of inference systems that minimize the total *cost* of errors rather than their total *numbers.* For instance, if a man must infer the sexual interest of a woman whom he encounters, he can make two types of errors: He can infer that she is sexually interested when she is not (false positive), or he can infer that she is not sexually interested when she is (false negative). What are the consequences of each type of error?[74]

The consequence of a false positive, thinking that she is interested when she is not, is that he would be turned down, maybe laughed at, possibly slapped in the face. The consequence of a false negative, thinking that she is not interested when she is, is a missed opportunity for sexual intercourse and to increase his reproductive success. The latter cost is far greater than the former. Thus, men should be selected to possess a cognitive bias that constantly leads them to overestimate a woman's sexual interest.

Haselton and Buss's Error Management Theory not only explains previously known phenomena, such as the results of the laboratory experiment mentioned above or the Safeway fiasco, but also leads to two novel predictions. First, women should underestimate a man's romantic commitment to them, because the cost of a false positive (thinking that a man is romantically committed to her when he is not, getting pregnant by him, and then having him desert her) is far greater than the cost of a false negative (thinking that he is not romantically committed to her when he is, and missing an opportunity to form a committed romantic relationship). If a woman misses an opportunity to form a long-term committed relationship with one man, she can soon get an opportunity to form one with another man; there are other fish in the sea. In contrast, one mistake with a wrong man can burden the woman with a child and ruin her future romantic prospects for many years to come.

Second, the tendency of men to overestimate a woman's sexual interest should not apply to their sisters' interest in other men, because men need to perceive their sisters' sexual interest in other men accurately so that they can protect them from unwelcome sexual advances from those they are not interested in. In other words, the cognitive bias of men to overestimate women's sexual interest is not blind or unqualified; it is only activated in encounters with women with whom they might conceivably have sex, which exclude their sisters. Haselton and Buss's studies confirm both of these novel predictions.

While Haselton and Buss apply their Error Management Theory exclusively to the area of mind reading between men and women (inference about sexual interest of potential mates), their insight may be applied to human behavior in other areas as well.[75] For example, evolutionary social psychologist Toshio Yamagishi and his colleagues suggest that people in social exchange situations make similar (and similarly unconscious) calculations when they decide whether or not to cooperate with each other.[76] They face the possibility of making two different types of errors: thinking that freeriding on others is possible without detection or punishment when it is not (false positive), or thinking that freeriding on others is not possible when it is (false negative). The cost of the former error is ostracism from the group, while the cost of the latter is the foregone benefit of exploiting others. Yamagishi and his colleagues suggest that people are less likely to commit the error of false positive when they are highly dependent on the group and cannot risk ostracism and expulsion from it.

As another example, Stewart Elliott Guthrie[77] and Pascal Boyer[78] use the same principle of cognitive bias to explain the emergence of religion. When something either good or bad happens to you, it might be the result of a purposeful, intentional act of someone (in other words, you have a friend or a foe that you may not know about), or it might be the result of a random event ("luck"). In deciding which it is, you can once again make two types of errors: thinking that it is a purposeful, intentional act when it is random (false positive), or thinking that it is random when it is a purposeful, intentional act (false negative). The consequence of the first type of error (ignoring a potential friend or foe) is far greater than the consequence of the second type of error (being a bit too paranoid or anthropomorphic). Guthrie and Boyer argue that religion and the human tendency to believe in God are byproducts of an evolved cognitive bias toward anthropomorphism. (See the section "Where Does Religion Come From?" in chapter 8 for further details.) In other words, at the most abstract level, it may be that we believe in God for the same reason that men constantly think that women are coming on to them: Because getting it wrong the other way would be much worse.

4

# Go Together Like a Horse and Carriage?

## THE EVOLUTIONARY PSYCHOLOGY OF MARRIAGE

It is a mistake to think that marriage is unique to the human species. While, of course, some of the specific accoutrements of human marriage–such as the wedding ceremony–is unique to humans, the institution of marriage itself–the predictable and regulated patterns of matings between a male and a female–is shared by many other species, particularly birds.[1] Further, some of the specifics of a Western marriage–the church wedding, marriage certificates–are not even human universals.

Because marriage is closely related to sex and mating, this is another area where evolutionary psychology has produced a large number of fascinating studies. Perhaps two of the most surprising findings of evolutionary psychology and biology (to be discussed in greater detail in this chapter) are about polygyny (marriage of one man to many women). First, despite the impression you might get from the history of Western civilization in the last millennium, humans are naturally polygynous, not monogamous, and as a result, *all* human societies (including the United States) are polygynous to various degrees. Second, contrary to what you might think, most women benefit from polygyny, while, conversely, most men benefit from monogamy.

Intrigued? Then read on. . . .

## Q. Why Are There Virtually No Polyandrous Societies?

First, let's get the terminology straight. As we discussed before, *monogamy* is the marriage of one man to one woman, *polygyny* is the marriage of one man to more than one woman, and *polyandry* is the marriage of one woman to more than one man. *Polygamy*, even though it is often used in common discourse as a synonym for polygyny, refers to both polygyny and polyandry. We will not use this ambiguous word in this book.

A comprehensive survey of traditional societies in the world shows that 83.39 percent of them practice polygyny, 16.14 percent practice monogamy, and 0.47 percent practice polyandry.[2] Almost all of the few polyandrous societies practice what anthropologists call *fraternal polyandry*, where a group of brothers share a wife. Nonfraternal polyandry, where a group of unrelated men share a wife, is virtually nonexistent in human society.[3] Why is nonfraternal polyandry so rare?

As we discussed in chapter 2, paternity certainty is low enough in a monogamous marriage, where the woman is "supposed to" be mating with only one man; the estimates of cuckoldry (where the man unknowingly raises another man's genetic child) *in monogamous societies* range from 13–20 percent in the United States, 10–14 percent in Mexico, and 9–17 percent in Germany.[4] When multiple men are officially married to one woman, who is "supposed to" mate with all of them, the co-husbands have very little reason to believe that a given child of hers is genetically his, and will therefore not be very motivated to invest in it. If the children receive insufficient paternal investment, they will not survive long enough to become adults and continue the society. Nonfraternal polyandry therefore contains the seeds of its own extinction.

In contrast, fraternal polyandry, where the co-husbands are brothers, can survive as a marriage institution because even when a given husband is not the genetic father of a given child (sharing half of his genes), he is at least the genetic uncle (sharing a quarter of his genes). The child of a fraternal polyandrous marriage could never be completely genetically unrelated to any of the co-husbands (assuming, of course, that the wife has not mated with anyone outside of the polyandrous marriage), so all the co-husbands are motivated to invest in all the children.

By the same token, the most successful type of polygyny is the sororal polygyny, where all the co-wives are sisters (although, unlike nonfraternal polyandry, nonsororal polygyny is very common). While a woman, when given a choice between marrying an unmarried man and marrying a married man, might under some circumstances rationally choose to marry polygynously (see the section "Why (and How) Are Contemporary Westerners *Polygynous*?" later in this chapter), it is never in the existing wife's material interest for her husband to acquire another wife. Every senior wife who is already married to the man suffers from the addition of each new wife to the household, because each additional wife takes away the husband's resources, otherwise available to her and her children. Thus, conflict among co-wives in polygynous marriages is very common, and for this reason polygynous men in many traditional societies maintain a separate household for each wife.[5] However, the conflict and competition for the limited resources of the husband are somewhat alleviated when the co-wives are sisters because then they will not object so strongly to the diversion of the resources to the new wife and her children, to whom the senior wife is genetically related.[6]

## *If You Want to Know What Women Have Been Up to, Look at Men's Genitals*

Now, the fact that polyandry is very rare in human society decidedly does *not* mean that married women have always been faithful to their husbands and mated with only one man. On the contrary, *human females have been promiscuous throughout their evolutionary history.* (Recall the dangers of moralistic fallacy from the introduction. The fact that marital fidelity is a virtue means neither that it is natural for us nor that we are always faithful to our spouses. Promiscuity may be morally good or bad, but its evolutionary naturalness has no bearing on the question.) How do we know? There are several pieces of evidence that support this conclusion. First, the very high rates of cuckoldry (men unwittingly raising another man's genetic children) in many contemporary societies cited above strongly suggest that extra-pair copulation (mating with sex partners to whom one is not formally married) has been an evolved strategy for females (both human and other species).[7]

Second, it turns out that we can measure the degree of female promiscuity rather precisely by *the relative size of testes on the male body.* Across species, the more promiscuous the females are, the larger the size of the testes relative to the male's body weight. This is because when a female copulates with multiple males within a short period of time–in other words, when they are promiscuous–sperm from different males must compete with each other to reach the egg to inseminate it. One good way to out-compete others is to outnumber them. Male gorillas, whose females live in a harem controlled by one silverback male and therefore do not have many opportunities for extra-pair copulations, have relatively small testes (0.02 percent of body weight) and produce a very small number of sperm per ejaculate ($5 \times 10^7$). On the other extreme, male chimpanzees, whose females are highly promiscuous and do not attach themselves to any single male, have relatively large testes (0.3 percent of body weight) and produce a very large number of sperm per ejaculate ($60 \times 10^7$).[8] On this scale, humans lie somewhere between the gorilla and the chimpanzee, but closer to the former than the latter. Men's testes are about 0.04–0.08 percent of their body weight, and the approximate number of sperm per ejaculate is $25 \times 10^7$. So human females have been more promiscuous than gorilla females in their evolutionary history, but not nearly as promiscuous as chimpanzee females. The evidence of women's promiscuity throughout evolutionary history is in the relative size of men's testicles. Men would not have such large testicles and produce so many sperm per ejaculate had women not been so promiscuous.

Finally, according to the pioneer biopsychologist Gordon G. Gallup and his collaborators, another piece of physiological evidence of promiscuity among human females in the evolutionary past is *the precise shape of the human penis.* The shape of the human penis is quite distinct from that of many other primate species. In particular, the glans ("head") of the human penis is shaped like a wedge. "The diameter of the posterior glans is larger than the penis shaft itself, and the coronal ridge, which rises at the interface between the glans and the shaft, is positioned perpendicular to the shaft."[9]

In addition, the human male during copulation engages in

repeated thrusting motions before he ejaculates. The combined effect of the particular shape of the penis glans and the repeated thrusting motions "would be to draw foreign semen back away from the cervix. . . . If a female copulated with more than one male within a short period of time, this would allow subsequent males to "scoop out" semen left by others before ejaculating."[10] In other words, the human penis is a "semen displacement device."[11] If human females did not engage in extensive extra-pair copulations throughout human evolutionary history, the human penis would not be shaped as it is, and the human male would not engage in repeated thrusting motions before ejaculating.[12] Clear evidence of women's promiscuity throughout evolutionary history is in the size and shape of men's genitals and what men do with them.

## **Q.** Why (and How) Are Contemporary Westerners *Polygynous*?

Polyandry (a marriage of one woman to many men) is very rare in human society (see "Why Are There Virtually No Polyandrous Societies?" above). This means that almost all human societies practice either monogamy or polygyny, which is the reason why the term *polygamy* is often used synonymously with *polygyny.* Polygyny is the only form of polygamy widely practiced in human societies, and a vast majority of human societies practice polygyny. Even though those of us in Western industrial societies tend to think of monogamy as both natural and normal, and even though Judeo-Christian religious traditions tell us that monogamy is the only natural form of marriage, monogamous societies are a small minority throughout the world. Why is this?

This is because, contrary to the Judeo-Christian tradition, humans are naturally polygynous.[13] By *naturally*, we mean that humans have been polygynous throughout most of their evolutionary history. (Recall the danger of naturalistic fallacy from our introduction. "Natural" means neither good nor desirable.) Strict and socially imposed monogamy is a recent invention in human evolutionary history. Unlike physical artifacts, however, human practices (like the institution of marriage) do not leave fossil records. How, then, do we know that our ancestors practiced polygyny more than ten thousand years ago in the ancestral environment?

It turns out that the clear evidence of our ancestors' polygyny is embodied in each of us. Both among primate and nonprimate species, the species-typical degree of polygyny (how polygynous members of a given species are on average) highly correlates with the degree of sexual dimorphism in size (the extent to which males of a species are larger than females).[14] The more polygynous the species, the greater the size disparity between the sexes. For example, among the completely monogamous gibbons, there is no sexual dimorphism in size; both by height and by weight males are about the same size as females. In contrast, among the extremely polygynous gorillas, males are 1.3 times as large by height and twice as large by weight as females.[15]

On this scale, humans are somewhere in the middle, but closer to the gibbons' end than that of the gorillas. Typically, human males are 1.1 times as large by height and 1.2 times as large by weight as human females.[16] This suggests that, throughout evolutionary history, humans have been *mildly* polygynous, not as polygynous as gorillas but not completely monogamous like gibbons either. This is how we know that humans are *naturally* polygynous.

### *Why Is Polygyny Related to Sex Differences in Body Size?*

However, this begs the question: Why does the degree of sexual dimorphism in size correlate with the degree of polygyny? There are two possible explanations of this correlation. The first, more established theory posits that males have become larger throughout evolutionary history; the second, newer theory argues that females have become smaller.

### *Did Men Become Bigger . . .*

The proponents of the first theory[17] point out that relative to monogamy, polygyny creates greater fitness variance (the distance between the "winners" and the "losers" in the reproductive game) among males than among females by allowing a few males to monopolize all females in the group. (See chapter 2, "Why Are Men and Women So Different?") The greater fitness variance among males creates greater pressure for men to compete with each other for mates. Under such severe physical competition, only big and tall males can emerge victorious and get mating opportunities, while small and short males are left out of the reproductive opportunities altogether. At the same time, among pair-bonding species like humans, where males and females stay together to raise their children, females prefer to mate with big and tall males who can provide better physical protection for themselves and for their children against predators and other males. Thus, through both competition among men and preference by women, only big and tall males can reproduce and pass on their "big and tall" genes to their sons, while most or all females of all sizes reproduce and pass on their full range of sizes to their daughters. (Remember, the "fitness floor"–the worst one can do–is relatively high for women.) Over many generations, males will get bigger and taller, while females will retain the same distributions of height and weight in each generation.

Recent critics[18] point out that this theory assumes that body size (height and weight) is transmitted exclusively or largely along the sex lines, from fathers to sons, and from mothers to daughters. The theory assumes that tall men married to short women have tall sons but short daughters. The critics use Finnish data on twins[19] to demonstrate that this assumption is false. The data show that sons are just as likely to inherit their height from mothers as from fathers, and daughters are just as likely to inherit their height from fathers as from mothers. So a tall father will have both tall sons and tall daughters, and a short mother will have both short sons and short daughters. What gives?

### *. . . or Did Women Become Smaller?*

The critics then point out that under polygyny, there is an evolutionary pressure for girls to mature earlier (see "Why Do Girls of Divorced Parents Experience Puberty Earlier Than Girls Whose Parents Remain Married?" in chapter 5). Under monogamy, most adult males are already married and cannot marry again, so there are no incentives for prepubescent girls to mature earlier; prepubescent boys in their age group are in no position to marry them. In contrast, under polygyny, married adult males can acquire additional wives. So girls who mature early can become a junior wife of a wealthy village chief while their prepubescent age mates cannot. Because girls who mature early attain smaller adult height than girls who mature late throughout the world (because girls essentially stop growing when they reach puberty),[20] this suggests that height differences between the sexes should be greater in polygynous societies as a result of girls undergoing earlier puberty and becoming shorter. Cross-cultural data show that this is indeed the case. Girls in polygynous societies are shorter than girls in monogamous societies, whereas boys from both types of societies are about the same height.[21]

Whichever theory is correct, it appears to be the case that polygyny and sex differences in height are closely related. This is how we know that humans are naturally polygynous: because men are taller than women.

### What Women Want

If humans are naturally polygynous, why, then, do many human societies in the world today practice monogamy (even though a large majority still practices polygyny)? One theory suggests that it is because that is what women want. In any species for which the female makes a greater investment in children than does the male (including humans), sex and mating is a female choice. Sexual intercourse occurs if and when the female wants it; the male has very little choice (outside of forcible rape).[22] (See "What Do Bill Gates and Paul McCartney Have in Common with Criminals?" in chapter 6.) Humans are no exception. Monogamy emerges as the institution of marriage in the society when many or most women choose to marry monogamously, and polygyny similarly emerges as the institution of marriage when many or most women choose to marry polygynously.[23]

What, then, would lead women to choose to marry monogamously or polygynously? One important determinant of the institution of marriage is the degree of resource inequality among men (the difference between the richest men and the poorest men). In societies with a high degree of resource inequality, where rich men are very much richer than poor men, women (and their children) are better off sharing the few wealthy men, because one-half, one-quarter, or even one-tenth of a wealthy man is still better than a whole of a poor man when resource inequality is extreme. Or, as George Bernard Shaw puts it, "the maternal instinct leads a woman to prefer a tenth share in a first rate man to the exclusive possession of a third rate one."[24]

In contrast, in societies with a low degree of resource inequality, where rich men are not much richer than poor men, women (and their children) are better off monopolizing a poor man than sharing a rich man, because one-half of a rich man will not be as good as a whole of a poor man.[25] Thus, polygyny emerges as the institution of marriage in societies characterized by greater resource inequality among men, while monogamy emerges in societies characterized by lesser resource inequality. This theory is an extension to human society of what is known as the polygyny threshold model in biology, originally formulated to explain the mating systems of birds,[26] thus once again illustrating the fundamental principle of evolutionary psychology that humans are no different from other species. (See "The Evolutionary Psychological Perspective" in chapter 1.)

The reason most Western industrial societies are monogamous, despite the fact that humans are naturally polygynous, is that men in such societies tend to be more or less equal in their resources, compared to their ancestors in medieval times. The degree of inequality tends to increase as societies become more complex, from hunter-gatherer and pastoral societies, to horticultural and agrarian societies, and typically reaches its maximum in advanced agrarian societies.[27] Industrialization tends to decrease the level of inequality in society.

Individual decisions of women to marry monogamously rather than polygynously combine to produce social institution and norms.[28] If many or most women choose to marry monogamously, then the society becomes monogamous. However, the true polygynous nature of humans is never too far beneath the surface, even in nominally monogamous societies such as ours.

### *All Human Societies Are Polygynous, Simultaneously or Serially*

Wealthy and powerful men throughout history, even while monogamously married, have always mated polygynously by having mistresses, concubines, and other extramarital affairs.[29] (See "What Do Bill Gates and Paul McCartney Have in Common with Criminals?" in chapter 6.) And it is true even today. Whether married or not, wealthier men in the United States and Canada have more sex partners and have sex more frequently than less wealthy men.[30] This is not because wealthy men can afford the services of prostitutes; wealthy men are no more likely to have sex with a prostitute than are poorer men. They do not have to. Wealthy men have more sex partners and have sex more frequently because women seek them out.

Most nominally monogamous societies also allow people to get a divorce, and in many societies, such as the United States, divorce is both very easy and very common. Liberal divorce laws allow men in these societies to practice *serial polygyny* (a man having multiple wives, not simultaneously but sequentially, through a series of divorce and remarriage). In the United States, the strongest predictor of remarriage after divorce is sex (male vs. female): men typically remarry, women typically do not. As we discuss in chapter 3, this is because men become more desirable with age to potential mates (thanks to the greater income and higher status that typically accompany age), while women become less desirable with age due to declining reproductive value and fertility. While some women do remarry after divorce and thus practice serial polyandry, a far greater number of men practice serial polygyny through divorce and remarriage. Contemporary Westerners who live in nominally monogamous societies that nonetheless permit divorce are therefore in effect polygynous; they practice serial polygyny.

## *Most Women Benefit from Polygyny, Most Men Benefit from Monogamy*

When there is resource inequality among men (which there always is in every human society), most women benefit from polygyny. This is because under polygyny, women can share a wealthy man, whereas under monogamy, they are stuck with marrying a poorer man. If the resource inequality is large enough, then a fraction of a wealthy man is bigger and thus better than a whole of a poor man.[31]

The only exceptions are extremely desirable women. These women can marry the most desirable, wealthiest men under any circumstance (polygyny or monogamy). Under monogamy, they can monopolize the wealthiest men, whereas under polygyny, they must share them with other, less desirable women. So the most desirable women benefit from monogamy, but all other women benefit from polygyny.

The situation is exactly opposite for men. Most men benefit from monogamy, because it guarantees that every man can find a wife. True, less desirable men can only marry less desirable women, but marrying a less desirable woman is much better than not marrying anyone at all.

Once again, extremely desirable men are the exceptions. Such men can have multiple wives under polygyny, whereas they are limited to only one wife (albeit an extremely desirable one) under monogamy. So extremely desirable men benefit from polygyny, but all other men benefit from monogamy.

When men in monogamous societies imagine what their life might be like under polygyny, they imagine themselves with multiple wives. So they may think they would be better off under polygyny. What they don't realize is that for most men, who are not extremely desirable, polygyny means no wife at all, or, if they are lucky, one wife who is much less desirable than one they could get under monogamy (because under polygyny, more desirable wives are taken by men who are more desirable than them). If they do the math, they will come to the right conclusion that most of them are better off under monogamy than under polygyny.

## Q. Why Does Having Sons Reduce the Likelihood of Divorce?

Sociologists and demographers have discovered that the presence of sons decreases the probability of divorce.[32] Couples who have at least one son face a significantly lower risk of divorce than couples who have only daughters. Why is this?

Remember from chapter 3 that a man's mate value is largely determined by his wealth, status, and power, whereas a woman's mate value is largely determined by her youth and physical attractiveness. This means that the father has to make sure that his son will inherit his wealth, status, and power, regardless of how much or how little of these resources he has. A working-class father still has to make sure that his son will inherit what little wealth he has, because the more the son inherits, the greater his expected reproductive success. In sharp contrast, there is relatively little that a father (or mother) can do to affect the daughter's expected reproductive success; once she is born, there is very little parents can do to keep her youthful or make her more physically attractive.

The evolutionary psychological logic therefore predicts that the continued presence of (and investment by) the father is important for the son, but not as much for the daughter. Strictly in reproductive terms, there is very little fathers (or anyone else) can do for daughters beyond keeping them alive and healthy. The presence of sons therefore deters divorce and departure of the father from the family more than the presence of daughters, and this effect should be stronger among wealthy families.

Of course, strongly wedded to the Standard Social Science Model as they are, the sociologists and demographers who discover that the presence of sons decreases the probability of divorce explain this finding by saying that fathers are *considered* to be more important for their sons' lives than for their daughters', and the presence of sons *encourages* fathers to get more involved in child rearing, thereby lowering the likelihood of divorce. Of course, they are right; fathers *are* generally considered more important for sons than for daughters, and the presence of sons *does* encourage fathers to get more involved. But the Standard Social Science Model cannot explain *why* this is so; evolutionary psychology can.

## Q. Why Are Diamonds a Girl's Best Friend?

Because women make disproportionately greater parental investment in children than men do, their primary task is to discriminate between "dads" and "cads"[33] among male suitors. *Dads* are males who are willing to invest in a woman and her offspring in the long run; *cads* are those who are only looking for cheap thrills for the night and are likely to desert her after having sex. Given that women can have only so many children in their lifetimes and that they must invest much more in each child, the reproductive consequences faced by a woman for failing to discriminate between dads and cads are very large.

How might a woman accomplish this task? How would she know which men will invest resources in her and her offspring? A good dad must possess two qualities: the *ability* to acquire and accumulate resources, and the *willingness* to invest them in her and her children. A good way to screen for men who are simultaneously able and willing to invest is to demand an expensive gift; only men who are capable of acquiring resources and willing to invest them can afford to give a woman expensive gifts, which are known as *courtship gifts* or *nuptial gifts* in evolutionary biology.[34] (Yes, females of other species demand these gifts before they agree to have sex with the males.) Would any expensive gifts do? A Mercedes-Benz? A house in the suburbs?

No, these gifts will not do. A man who is *intrinsically* interested in luxury European cars might buy her a Mercedes. A man who is *intrinsically* interested in real estate might buy her a house in the suburbs. In either case, his gift is not an unequivocal and pure indicator of his *general* and *universal* willingness to invest resources in her and her offspring. The courtship gift for the purpose of screening dads from cads must not only be costly but also lack intrinsic value.

Diamonds make excellent courtship gifts from this perspective because they are simultaneously very expensive and lack intrinsic value. No man (or woman) can be inherently interested in diamonds; you cannot drive them, you cannot live in them, you cannot do *anything* with them. Any man who would buy diamonds for a woman must be interested in making an investment in her. Flowers, another favored gift for women, are also relatively expensive and lack intrinsic value. Of course, diamonds and flowers are beautiful, but they are beautiful precisely because they are expensive and lack intrinsic value, which is why it is mostly women who think flowers and diamonds are beautiful. Their beauty lies in their inherent uselessness; this is why Volvos and potatoes are not beautiful.

Consistent with this evolutionary psychological logic, a recent analysis using game theory demonstrates that what the researchers call "extravagant" gifts–gifts to women that are "costly but worthless"–facilitate courtship.[35] The researchers note that such extravagant gifts have the added benefit *for men* of deterring "gold diggers," women who promise to mate in exchange for a gift but then desert without mating after receiving it. (Once again, yes, there are such "gold diggers" among other species as well.) It appears that women are not the only ones who must screen their mates very carefully.

## Q. Why Might Handsome Men Make Bad Husbands?

Recall from chapter 3 that beauty is *not* in the eye of the beholder or skin-deep. Beautiful people are genetically and developmentally healthier than are unattractive people. So how are handsome men different from unattractive men as husbands? Why would healthier men not make better husbands?

Two leading evolutionary psychologists, Steven W. Gangestad and Jeffry A. Simpson, suggest one answer.[36] Gangestad and Simpson observe that men can maximize their reproductive success by pursuing one of two different strategies: Seek a long-term mate, stay with her, and invest in their joint offspring (the "dad" strategy); or seek a large number of short-term mates without investing in any of the resulting offspring (the "cad" strategy).[37]

All men may want to pursue the cad strategy; however, their choice of the mating strategy is constrained by female choice. Men do not get to decide with whom to have sex; women do.[38] And women disproportionately seek out handsome men for their short-term mates for their good genes. Even women who are already married would benefit from short-term mating with handsome men if they could successfully fool their husbands into investing in the resulting offspring. The women then get the best of both worlds: Their children carry the high-quality genes of their handsome lover and the parental investment of their unknowingly cuckolded but resourceful husband.

Thus, handsome men get a disproportionate number of opportunities for short-term mating and are therefore able to engage in the cad strategy. Ugly men have no choice. Since women do not choose them as short-term mates, their only option for achieving any reproductive success is to find one long-term mate and invest heavily in their children—the dad strategy.

Consistent with Gangestad and Simpson's theory, a study shows that more attractive men have a larger number of extra-pair sex partners (sex partners other than their long-term mates).[39] Another study shows that more attractive men have more short-term mates than long-term mates, while more attractive women have more long-term mates than short-term mates.[40] More important, handsome men invest less in their exclusive relationships than ugly men. They are less honest with and less attentive to their partners.[41]

We hasten to add that "good" and "bad" are value judgments that we promised not to make in this book. (See "Two Errors in Thinking That We Must Avoid" in the introduction.) However, empirical data do demonstrate that handsome men have more extramarital affairs and are not as committed to their marriages, which many wives may consider undesirable. In this sense, handsome men make better lovers than husbands.

# 5

# Some Things Are More Important Than Money

## THE EVOLUTIONARY PSYCHOLOGY OF THE FAMILY

While the topic of the family receives somewhat less attention in evolutionary psychology than, say, sex and mating do, evolutionary psychologists have nonetheless made significant contributions toward our understanding of the human family. Two of the early pioneers of modern evolutionary psychology, Martin Daly and Margo Wilson, for example, conducted a study in Canada and the United States demonstrating the dangers of stepparents–stepfathers in particular–to children.[1] Infants and children who do not live with two biological parents face 40 to 100 times as great a chance of being injured or killed within the family as those who live with both biological parents. In a sense, Daly and Wilson provided an evolutionary psychological explanation for the "Cinderella Effect."[2]

Family is the context for parental investment; it is where children are born and raised by biological and not-so-biological parents (who believe themselves to be biological parents but really are not). While much of parental investment in children is made consciously by the parents, one of the great and surprising discoveries of evolutionary psychology is that some forms of parental investment are unconsciously made.

For example, parents may invest more or less into sons or daughters simply by having more children of one sex over the other. The sex of the child is not consciously decided by the parents (outside of sex-selective abortion). While parents may wish to have a boy or a girl, they cannot consciously choose to have one or the other. Yet evidence shows that the sex of the child can be predicted by certain features of the parents that are important in evolutionary psychological terms.

That's where our story begins. . . .

## **Q.** Boy or Girl? What Influences the Sex of Your Child?

It is commonly believed that whether parents conceive a boy or a girl is *entirely* up to chance. Close enough, but not quite; it is *largely* up to chance, but there are factors that very subtly influence the sex of an offspring. It is also commonly believed that exactly half the babies born are boys and the other half are girls. Close enough, but not quite; the normal sex ratio at birth is 0.5122–that is, 105 boys born for every 100 girls. But the sex ratio varies slightly in different circumstances and for different families. So what factors affect the sex of the child?

### *The Genius of Robert L. Trivers*

Any discussion of sex ratio at birth must begin with the work of Robert L. Trivers, who is one of the greatest evolutionary biologists of our time. In 1973, Trivers teamed up with a mathematician, Dan E. Willard, to formulate one of the most celebrated principles in evolutionary biology, called the Trivers-Willard hypothesis.[3] The hypothesis states that wealthy parents of high status have more sons, while poor parents of low status have more daughters. This is because children generally inherit the wealth and social status of their parents. Sons from wealthy families, who themselves become wealthy, have, throughout most of evolutionary history, been able to expect to have a large number of wives, mistresses, and concubines, and produce dozens or hundreds of children,[4] whereas their equally wealthy sisters can have only so many children. So wealthy parents should "bet" on sons rather than daughters.

Conversely, poor sons can expect to be completely excluded from the reproductive game, because no women would choose them as their mates. But their equally poor sisters can still expect to have some children if they are young and beautiful. (Recall from chapter 2 that the "fitness ceiling"–the best one can do–is much higher for men than for women, while the "fitness floor"–the worst one can do–is much higher for women than for men.) So natural selection designs parents to have a biased sex ratio at birth depending upon their economic circumstances–more boys if they are wealthy, more girls if they are poor.

There is evidence for this hypothesis throughout human societies. American Presidents, Vice Presidents, and cabinet secretaries have more sons than daughters.[5] Poor Mukogodo herders in East Africa have more daughters than sons, both at birth and in the zero to four age group.[6] Church parish records from the seventeenth and eighteenth centuries in Germany show that wealthy landowners in Leezen, Schleswig-Holstein, had more sons than daughters, while farm laborers and tradesmen without property had more daughters than sons.[7] Among the Cheyenne Indians on the American Plains, prestigious, high-status "peace chiefs" have more sons than daughters, while poor and marginal "war chiefs" have more daughters than sons in the zero to four age group.[8] In the contemporary United States and Germany, the elite–judged by the listing in their respective country's *Who's Who*–have a greater proportion of sons among their offspring than does the population in general.[9] In an international survey of a large number of respondents from forty-six different nations, more wealthy individuals are more likely to indicate a preference for sons if they could only have one child, whereas less wealthy individuals are more likely to indicate a preference for daughters.[10] While there is some counterevidence,[11] most evidence is in support of the Trivers-Willard hypothesis.[12]

### *Extending Trivers's Genius*

Recently, there has been a theoretical extension of the original Trivers-Willard hypothesis, called the *generalized* Trivers-Willard hypothesis.[13] The idea behind the new hypothesis is the same as that behind the old one, but it extends the idea to many other factors besides the family's wealth and status. The new hypothesis suggests that if parents have any trait they can pass on to their children that is better for sons than for daughters, then they will have more boys. Conversely, if parents have any trait they can pass on to their children that is better for daughters than for sons, then they will have more girls. Parental wealth and status are just two of the traits they can pass on to their children that are more beneficial for sons than for daughters, but there are many other factors.

Brain types are another example of such heritable traits. Strong "male brains," which are good at systematizing (figuring things out), are more beneficial for sons than for daughters, while strong "female brains," which are good at empathizing (relating to people), are more beneficial for daughters than for sons.[14] Since brain types are heritable, the generalized Trivers-Willard hypothesis would predict that parents with strong male brains, such as engineers, mathematicians, and scientists, are more likely to have sons, while those with strong female brains, such as nurses, social workers, and school teachers, are more likely to have daughters. This is indeed the case.[15] While the sex ratio at birth among the general population is 0.5122–that is, 105 boys for every 100 girls–the study shows that the sex ratio among engineers and other systemizers is 0.5833–that is, 140 boys for every 100 girls. The comparable sex ratio among nurses and other empathizers is 0.4255–that is, 140 girls for every 100 boys.[16]

By the same token, tall and big parents have more sons and produce more male fetuses (because size was a distinct advantage in male competition for mates in the ancestral environment, while body size has no particular advantage for women), and short and small parents have more daughters and produce more female fetuses.[17] Because violence was probably a routine means in the male competition for mates in the ancestral environment[18] (as it is among our primate cousins),[19] tendency toward violence was adaptive for ancestral men but not for ancestral women. Accordingly, violent men have more sons, both in the United States and the United Kingdom.[20]

## *Why Beautiful People Have More Daughters . . .*

Physical attractiveness can also bias the sex of your children. Now, unlike being big and tall or having a tendency toward violence, which increases the reproductive success of only men and not women, being beautiful is good for both men and women. Beautiful women have greater mating success than less attractive women, and handsome men do better than less attractive men. But beautiful men and beautiful women tend to do "better" in slightly different ways.

Physically attractive women tend to do well both in long-term and short-term mating; men prefer beautiful women for both. In contrast, handsome men tend to do well mostly in short-term mating. Women seek out handsome men for short-term mating (possibly to get good genes for their children by being impregnated by them but then passing the resulting offspring off as that of their unsuspecting husband) but not necessarily for long-term mating, for which other qualities like the man's resources and status become more important. In fact, as we suggest in chapter 4 (see "Why Might Handsome Men Make Bad Husbands?"), physically attractive men may not make desirable long-term mates for many reasons.

So physical attractiveness, while a universally positive quality, contributes even more to women's reproductive success than to men's. The new hypothesis would therefore predict that physically attractive parents should have more daughters than sons. Once again, this is indeed the case. Young Americans who are rated "very attractive" have a 44 percent chance of having a son for their first child (and thus a 56 percent chance of having a daughter). In contrast, everyone else has a 52 percent chance of having a son (and thus a 48 percent chance of having a daughter) for their first child.[21] Being "very attractive" increases the odds of having a daughter by 36 percent!

## *. . . and Women Are More Beautiful Than Men*

If you look around and rate the men and women around you on their physical attractiveness, you should notice that, whether you are a man or a woman, gay or straight, women on average are objectively more attractive than men. Why might this be the case?

Think about it. *If* physical attractiveness is heritable, such that beautiful parents beget beautiful children (and less attractive parents beget similarly less attractive children), *and if* beautiful people are more likely to have daughters than sons, then it logically follows that over time, women will become more attractive on average than men. Once again, studies confirm this implication of the new hypothesis.[22] The average level of physical attractiveness among women is significantly higher than the average level of physical attractiveness among men. Women are more beautiful than men because beautiful parents have more daughters than sons.

Far from being random chance, there are a large number of factors that appear to influence, even if only very slightly, whether a couple will have a son or a daughter. The generalized Trivers-Willard hypothesis can explain the evolutionary reasons why these factors affect the sex of the child. Evolution helps parents pass on their genes in the most efficient way possible.

## Q. Why Does the Baby Have Daddy's Eyes but Not Mommy's?

As we discussed in chapter 2, because of the sexual asymmetries in reproductive biology, the possibility of cuckoldry exists only for men. Men can be cuckolded and unwittingly invest their limited resources in someone else's genetic children, whereas women could never be cuckolded. In other words, paternity can never be certain, while maternity is always certain. This is well expressed in the common saying "Mommy's baby, Daddy's maybe."

Men who are cuckolded do not manage to transmit their genes to the next generation and achieve no reproductive success. Men are therefore selected to be very sensitive to cues to possible cuckoldry and to attempt to guard against the possibility. A man would therefore only invest in his mate's children if he was reasonably certain that they were genetically his. In the absence of DNA tests (which did not exist in the ancestral environment), how could men ever be certain that their children were genetically theirs?

The child's physical resemblance would be one clue available to men in the ancestral environment. If the baby looks like the father, it is more likely that it is genetically his, whereas if the baby looks nothing like him, or, worse yet, looks a lot like his neighbor, then it is doubtful that he is its genetic father. This reasoning leads evolutionary psychologists to predict that, holding constant the probability of cuckoldry, babies who resemble their father are more likely to survive than babies who do not resemble him (or resemble the mother), because the father of babies who resemble him is more likely to be convinced of his paternity and to invest in them, thereby increasing their chances of survival, whereas the father of babies who do not resemble him (or resemble the mother) is less likely to be convinced of his paternity and to invest in them, thereby decreasing their chances of survival. Over many generations throughout evolutionary history, genes that make babies resemble the father therefore survive, whereas genes that make them resemble the mother do not, and so more and more babies come to resemble the father, until most babies are born resembling the father, not the mother.

This is precisely what two psychologists at the University of California, San Diego, Nicholas J. S. Christenfeld and Emily A. Hill, discover in their ingenious study.[23] Christenfeld and Hill show the subjects in their experiment a picture of a child at ages 1, 10, and 20, and a set of three pictures of adults, one of whom is the real parent (mother or father) of the child. They then ask the subjects to match the child with the correct parent. Christenfeld and Hill's subjects therefore have 0.33 probability of selecting the right parent by chance.

If the child truly resembles the parent, then the subjects should be able to match the two pictures at a much higher probability.

A major finding in Christenfeld and Hill's experiment is that children in general do not physically resemble their parents. The subjects are not able to match the picture of the child at any age to the picture of either the mother or the father better than expected by chance. The only exception, however, is the matching of 1-year-old babies to their father. The subjects are able to match both baby boys (0.505) and baby girls (0.480) to their father (though not to their mother) at statistically significantly greater rates than by chance. That means that one-year-old babies resemble their fathers, as might be expected from the evolutionary psychological logic presented above.

Christenfeld and Hill's finding was widely reported in the media, but it has also become one of the most controversial contentions in evolutionary psychology, not least because, although their explanation had impeccable logic, their finding could not be replicated. To date, attempts at replication have shown that newborn babies objectively resemble mothers more than fathers,[24] and infants and children resemble both parents equally.[25] Thus, the question of whether newborn babies objectively resemble the father more than the mother must be treated as an open one until more experiments are conducted.

## *But Who Are Newborn Babies* Said *to Resemble?*

There is a related finding that is much less controversial and well replicated, however. Nature may or may not help assure fathers of their paternity by making babies resemble them; however, friends and family–in particular, mothers and their kin–certainly do. In three separate studies conducted in three different North American countries (Canada, Mexico, and the United States) in three different decades, mothers and maternal relatives are far more likely to allege the baby's paternal resemblance than its maternal resemblance.[26] This happens even when the newborn babies in fact do *not* resemble their fathers.[27] Such allegations of paternal resemblance assure the fathers of their paternity, whether the babies actually resemble them or not.

Further, in most societies, babies get their last name from the father, not the mother, thereby once again suggesting to the father that he is the father of the baby. (Russians go one step further and give their babies their *middle and last names* after the father.) This is true even in societies where women routinely keep their last name when they get married and do not adopt their husband's name. The children of such parents nonetheless usually get their last name from the father, not the mother. Many Western professional women these days often keep their last name when they get married. Most of their children still get their last names from the father, not the mother. By giving their children the father's last name, these women are essentially (albeit unconsciously) saying, "Honey, it's yours" (even, or *especially*, when it is not). They need to reassure their husbands of their paternity, but do not themselves need to be reassured of their maternity; they know for sure. Mommy's baby, Daddy's maybe. And for 10–30 percent of daddies, it is not.

## **Q.** Why Are There So Many Deadbeat Dads but So Few Deadbeat Moms?

When married couples with children get divorced, chances are that the children stay with the mother, not the father, especially when they are young. According to the 1992 March/April Current Population

Survey in the United States, conducted by the US Census Bureau on a nationally representative sample, 86 percent of custodial parents are mothers.[28] Further, many of the noncustodial fathers who have agreed, either voluntarily or via court order, to pay child support default on their commitment and often become "deadbeat dads." The first national survey of the receipt of child support, conducted in 1978, reveals that less than half (49 percent) of women awarded child support actually receive the full amount due to them, and more than a quarter (28 percent) of them receive nothing.[29] The percentages have remained more or less constant since. In 1991, 52 percent of custodial parents awarded child support received the full amount; 25 percent of them received nothing.[30] Why are women so much more dedicated parents than men? Why are there so many deadbeat dads but so few deadbeat moms?

On the surface, this massive sex difference in the dedication to parenthood might appear puzzling, since both the mother and the father are equally related to their children genetically; each transmits half of their genes to their child. However, there are two biological factors that combine to make fathers far less committed as parents than mothers.

## *"Mommy's Baby, Daddy's Maybe"*

The first is paternity uncertainty. As we discussed in chapter 2 and above in this chapter ("Why Does the Baby Have Daddy's Eyes but Not Mommy's?"), because gestation for all mammals (including humans) takes place internally within the female's body, the male can never be certain of his paternity, whereas maternity is always certain. And paternity uncertainty is not a remote theoretical possibility. As we mention elsewhere (see "The Evolutionary Psychological Perspective" in chapter 1 and "Why Are There Virtually No Polyandrous Societies?" in chapter 4), the estimated incidences of cuckoldry (men unwittingly investing in another man's genetic offspring) in contemporary industrialized societies is substantial (between 10 and 30 percent), although a comprehensive recent review suggests that the actual incidence among Western populations may be much lower, at around 4 percent.[31] Thus, this is a very realistic possibility for any father in contemporary Western society (and probably elsewhere throughout history as well). Naturally, men are not motivated to invest in children who have a distinct possibility of not being genetically theirs.

### *The Best They Can Do Is Better for Men*

The second biological factor that makes fathers less committed parents is their higher fitness ceiling (the best they can do reproductively). Fetuses gestate for nine months within the female body, and infants are, at least in the past, nursed by the mother for several years after birth, during which the mother is usually infertile. Women also have a much shorter reproductive life than men do. These two factors combine to create a much higher fitness ceiling for men than for women. Men can potentially have many more children than women can. (Remember Moulay Ismail the Bloodthirsty?) The sex difference in the largest possible number of children means that, while *reproductive success* is equally important to men and women (in fact, to all biological organisms), *each child* is far more important to the mother than it is to the father. Each child represents a far greater portion of a woman's lifetime reproductive potential than it does a man's. If a 40-year-old mother of five deserts her children and they die as a result, she will likely end her life as a total reproductive loser, having failed to leave any copy of her genes in the next generation. If a 40-year-old father of five does the same, he can go on to produce five (or ten or twenty) more children.

Both paternity uncertainty and the higher fitness ceiling make fathers less committed parents than mothers, and this is why there are so many more deadbeat dads than deadbeat moms; very few women abandon or neglect their children. Ironically, it is the mother's greater commitment to her children that allows the father to neglect them even more.[32] Knowing the mother's greater commitment to her children, the father can abandon them, secure in the knowledge that the mother would never do likewise, because if she did, the children would be virtually certain to die. In other words, divorced parents with children are playing a game of chicken, and it is usually the mother who swerves. Most fathers would probably prefer to invest in their children and raise them by themselves rather than see them die, but they normally do not have to make this difficult decision, because they know that the mother would never abandon them. The mother's greater commitment to her children ironically allows the father to have his cake and eat it too, by moving on to the next marriage and family in which to invest.

## *Are Mothers Always Good Parents?*

None of this means that all mothers are always good parents or better parents than fathers. Sometimes mothers even kill their babies. However, evolutionary psychological logic can even explain who is more likely to kill their babies, and why.

Statistics show that very young mothers, by far, are the most likely to kill their babies, and older mothers are the second most likely to do so, but for different reasons.[33] Very young, teenage mothers kill their babies because they still have most of their reproductive lives ahead of them, and they can make more babies in the future even if they kill the one they just had. Having a baby under unfortunate circumstances (such as without the father willing to invest in it) not only threatens the well-being of the baby but also jeopardizes the mother's chance of finding a mate in the future. And teenage mothers are more likely to have their baby under unfortunate circumstances than others.

Older mothers (above the age of 35) kill their babies for a different reason. They are more likely to have defective babies because of their age. Every child (defective or otherwise) consumes parents' resources. Since defective children are much less likely to attain reproductive success, from the purely genetic point of view, any resources invested in children who will not have children themselves are wasted. Such children are taking away valuable resources from other children who have better reproductive prospects. Older mothers are more likely than younger mothers to have other children they must also raise. So parents are designed not to invest in defective children. (By the same token, parents invest more in better-looking children than in less good-looking children, and in more intelligent children than in less intelligent children.)

Yes, the evolutionary logic is very brutal, cold, and heartless. It only cares about the survival of the genes.

## Q. Why Is Family More Important to Women Than to Men?

Ask a group of friends, colleagues, and acquaintances (both men and women) to name five of their closest associates. Who are the people they talk to when they have something important to discuss? Chances are that women in your circles mention more family members among their closest associates, whereas men mention more coworkers and business associates in their personal networks. Studies in social networks repeatedly find that while otherwise comparable men and women have similar personal networks, women typically have more kin and fewer coworkers than men do.[34] Why is this? Why are women closer to their family members than men are?

Two sociologists, Lynn Smith-Lovin and J. Miller McPherson, propose an explanation for this universal phenomenon from the Standard Social Science Model perspective.[35] Using fictitious characters named Jim and Jane, they explain how the compositions of their personal networks remain more or less the same through adult years because "Jim is serious about his career as an engineer [and] Jane is equally serious about her nursing." However, the change begins when they become parents. "When their first child is born, however, Jane's mother comes to visit for two weeks; Jane begins to use her sister as a babysitter for daytime care while she is working. . . . Because more of her time is taken up with the baby, Jane's networks become more centered on neighborhood and kin, to some extent at the expense of her work and voluntary association friends. Jim's work and group ties are less altered."[36]

Jane and Jim in Smith-Lovin and McPherson's description accurately mirror what happens to many young couples when they have children. So, in that sense, their explanation is correct. However, it simply begs the question: Why is it Jane's mother, not Jim's, who comes to visit after the baby is born, when Jim's mother is presumably as equally related to the baby as Jane's mother? *Or is she?* Why is it Jane's sister, not Jim's, who becomes their babysitter, when both sisters are equally related to the baby? *Or are they?* Smith-Lovin and McPherson assume that it is Jane, not Jim, who is the primary caretaker of the baby. Why is this so? Their Standard Social Science Model explanation cannot answer these fundamental questions.

(We will return to another fundamental question of why Jim is an engineer and Jane is a nurse in chapter 7, "Why Are Most Neurosurgeons Male and Most Kindergarten Teachers Female?")

Evolutionary psychology can answer all of these questions. We have already addressed why mothers (like Jane) are more committed to their children than fathers (like Jim) are. (See "Why Are There So Many Deadbeat Dads but So Few Deadbeat Moms?" above.) So we know why Jane becomes the primary caretaker of the baby, not Jim (who may or may not be the baby's genetic father and who, if he becomes a successful engineer, might leave Jane and their child when Jane is 40 and marry the 20-year-old receptionist he hired for his engineering firm and start a new family anyway). Even though women are more motivated to make parental investment than men are, they cannot always do it alone; sometimes they need help from others. This was especially true back in the ancestral environment, where resources were scarce and life was precarious.

When mothers need help in their effort to raise their children, nobody is more likely or willing to deliver it than their kin. The mother's kin are more motivated to invest in the children than the father's kin, because the mother's kin know for sure that they are related to the children, whereas the father's kin may or may not be, due to paternity uncertainty. This is why it is Jane's mother, not Jim's, who comes to visit for two weeks, and it is Jane's sister, not Jim's, who babysits during the day. Jane's mother and sister are certain to be related to the child; Jim's mother and sister are not. This is why women today have a larger number of kin in their personal networks than men do. Women need to rely on their kin in case they need help, materially or otherwise.

Two implications follow from this logic. First, if women maintain their ties to their kin in case they need help with their parental investment, then women who are materially better off should need less help from their kin, and therefore have less need to maintain their ties with them. Second, women who are currently married should need less help from their kin than women without husbands, because even with residual paternity uncertainty, the putative fathers should be motivated to make some parental investment in the children and thereby lessen the mothers' burden. The presence of the spouse should be of at least some help. Further, the logic implies that family income and being currently married should have no effect on the extent to which men have kin in their personal networks.

An analysis of a large American data set confirms both of these predictions. A woman's family income and her currently being married both significantly decrease the proportion of kin in her personal network, whereas a man's family income and marital status have no effect on the proportion of kin in his network.[37] The Standard Social Science Model explanation cannot account for these patterns.

## Q. Why Do Girls of Divorced Parents Experience Puberty Earlier Than Girls Whose Parents Remain Married?

Developmental psychologists have known for nearly two decades that girls whose parents divorce early in their lives, particularly before the age of five, experience puberty earlier than their counterparts whose parents stay married.[38] Such girls are also more likely to start having sex at an earlier age, have a larger number of sex partners, get pregnant while still a teenager, and experience divorce in their first marriage.[39] Since the biological purpose of puberty is to mark the onset of the reproductive career, it makes perfect evolutionary sense that girls who undergo puberty at an earlier age start having sexual intercourse, have more sex partners, and get pregnant at an earlier age. (Recall once again the dangers of naturalistic fallacy. Just because something makes perfect evolutionary sense does not mean it is good or desirable.) But why does the presence or absence of the father at home during early childhood affect the age of puberty and thus the onset and promiscuity of sexual activity?

There are two competing explanations. One is that girls who experience puberty early are genetically different from those who experience it late. The other explanation is that girls have similar genetic makeup but respond to the environment differently, by starting puberty early or late. So which model is correct?

In the case of pubertal timing, both models are likely to be partially correct.[40] In support of the genetic model, there is substantial evidence that a girl's pubertal timing is largely heritable; about 50–80 percent of its variance is explained by genetic differences.[41] In this model, girls who undergo puberty early are *simultaneously* more likely to get a divorce because of their greater tendency toward sexual promiscuity, *and* to pass on their early puberty–greater promiscuity genes to their daughters. Hence, girls who grow up without a father (because their mother got a divorce or was never married) are more likely to experience puberty early and to become more sexually promiscuous because they have inherited the genes that will predispose them to do so.

While evidence supports the genetic model, environmental influences can also affect the actual timing of puberty within a window set by the genes.[42] This phenomenon is similar to other biological traits, such as height, weight, or intelligence. Height, for example, is highly heritable, so children of tall parents on average become taller than children of shorter parents; genes set the boundaries of potential adult height. Within these boundaries, however, environmental influences, such as nutrition or childhood exposure to disease, can determine the actual adult height.

Between 20 and 50 percent of pubertal timing is unaccounted for by the genes, so environmental conditions can still influence the actual onset of puberty within the windows set by the genes. One of the most important early childhood influences is the absence of the father.[43] In this model, girls who grow up without a father learn that men do not form lasting relationships with women and invest in their offspring. These girls then adopt a more promiscuous reproductive strategy of undergoing puberty early and forming short-term relationships with a large number of sex partners because they sense they cannot rely on men to form a committed relationship with them and provide parental investment in their offspring. In contrast, girls who grow up with a father at home learn the opposite lesson–that men do form lasting relationships with women and invest in their offspring. These girls then adopt a more restrained reproductive strategy of delaying their puberty and forming a committed long-term relationship with a partner who will invest in their offspring.[44] Hence, the presence or absence of a father in a girl's home before the age of five can explain both her age of puberty and her reproductive strategy.

## *Hold On . . .*

There is a piece missing from this explanation, however. In order for this strategy to evolve among women, men's tendency toward forming committed relationships and making parental investment must be stable across generations; the mother's experience with her mates must be highly predictive of the daughter's experience one generation later. One suggestion is that girls use the presence or absence of the father in the home as an indicator of the institution of marriage in the society.[45] In this view, father absence signifies not the man's unwillingness to form long-term relationships but a high degree of polygyny in society. In a highly polygynous society, married men are spread thinly among their multiple wives, and they cannot spend much time with any one of their wives or their offspring. Thus, the more polygynous the society, the less time any girl (or boy) spends with the father. In contrast, in monogamous societies, married men have only one wife, so they can spend all of their time with their wife and children. So the degree of father absence might be a microlevel indicator (within the family) of a macrolevel degree of polygyny (within the society).

In polygynous societies, there is an incentive for girls to mature early because any pubescent girl can become a junior wife of a wealthy polygynist while a prepubescent girl cannot. In contrast, there is no incentive for girls to mature early in monogamous societies because all adult men in such societies are already married (and cannot marry again), given a 50–50 sex ratio, and pubescent girls can only marry young teenage boys, who do not have the wealth or status to support a family. Consistent with this logic, an analysis of cross-cultural data shows that girls undergo puberty significantly earlier in polygynous societies and in nominally monogamous societies with a high incidence of divorce (and thus a higher incidence of serial polygyny; see "Why [and How] Are Contemporary Westerners *Polygynous*?" in chapter 4).[46] From this perspective, the average age of puberty has dropped precipitously in the United States in recent decades[47] because the divorce rate (and thus the incidence of remarriage for men, that is, serial polygyny) has increased dramatically.

The biochemical mechanism by which parental divorce

precipitates early puberty in girls is not well known. The evolutionary developmental psychologist Bruce J. Ellis[48] suggests that pheromones (a chemical substance that travels from one individual to another in order for the former to influence the behavior of the latter) emitted by the stepfather and other unrelated men in the household might trigger early puberty in girls. This is one of the remaining mysteries in evolutionary psychology.

# 6

# Guys Gone Wild

## THE EVOLUTIONARY PSYCHOLOGY OF CRIME AND VIOLENCE

While there have not been many evolutionary psychological studies in the area of crime and violence, one of the early classics in the field was on this topic.[1] Martin Daly and Margo Wilson's 1988 book *Homicide* demonstrates that all types of homicide–killing children, killing parents, men killing other men, men killing women, husbands killing wives, and wives killing husbands–can be explained by Darwinian logic.

At first sight, it appears that killing children makes no sense from the evolutionary psychological perspective, which emphasizes reproductive success. Why would parents kill their own children? Daly and Wilson have two answers to this question. The first answer is that they don't. Daly and Wilson discovered that what often passes as parents killing their children in police statistics is actually stepfathers killing their stepchildren, who do not carry their genes. It looks as though biological parents are killing their genetic children in the statistics because the police, uninformed by Darwinian logic, make no distinction between biological parents and stepparents in their record keeping. Biological parents very seldom kill their genetic children.

Their second answer to the question is that sometimes parents have to make tough choices. All parents, even wealthy ones, have limited resources to invest in their children. Every dollar, every minute, every effort that they invest in one child is another dollar, another minute, another effort that they cannot invest in other children. Their evolved psychological mechanisms therefore compel them to invest most efficiently, which usually means that they invest more in children who have the greatest prospect for reproductive success, at the cost of other children whose reproductive prospect is gloomier.

Another surprising finding in this area is that criminals are not so different from other men. All men (criminal or not) are more or less the same. The ultimate reason why men do what they do—whether they be criminals, musicians, painters, writers, or scientists—is to impress women so that they will sleep with them. Men do everything they do in order to get laid.

Of course, by the same token, all women are more or less the same, but you cannot see that in this chapter on crime and violence, because almost all criminals are men. Let's first find out why that is the case. . . .

## Q. Why Are Almost All Violent Criminals Men?

There are many *cultural universals*—features of human society that are shared by all known cultures. Donald E. Brown provided the original list of "human universals" (and wrote a whole book about them) in 1991,[2] and Steven Pinker updated the list in 2002.[3] There are probably so many cultural universals (contrary to what Franz Boas and cultural determinists think) because human culture is a manifestation of human nature at the level of society, and human nature is universal to all humans.[4] This is why all human cultures are more or less the same, and there are so many cultural universals. (See "What about Culture? Is Anything Cultural?" in chapter 2.)

Among the many cultural universals is the fact that men in every human society commit an overwhelming majority of all crimes and acts of violence.[5] Why is this? Why are men so much more criminal and violent than women?

In their comprehensive study of homicide, the leading evolutionary psychologists Martin Daly and Margo Wilson note that humans throughout their evolutionary history were effectively polygynous—many married men had multiple wives.[6] (See "Why [and How] Are Contemporary Westerners *Polygynous*?" in chapter 4.) In a polygynous society, some males monopolize reproductive access to all females while other males are left out; in such a society, some males do not get to reproduce at all, while almost all females do. (Recall our discussion of sex differences in fitness variance in chapter 2. The distance between the "winners" and the "losers" in the reproductive game is much greater among men than among women.) This inequality of reproductive success between males and females makes males highly competitive in their effort not to be left out of the reproductive game. This competition among men leads to a high level of violence (murder, assault, and battery) among them, and the large number of homicides between men (compared to the number of homicides between women, or between the sexes) is a direct result of this male competition for mates.

## *Big Violence Starts Small*

In particular, Daly and Wilson[7] note that most homicides between men originate from what is known as "trivial altercations."[8] A typical homicide in real life is not one depicted in an episode of *Columbo*: premeditated, well planned, and nearly perfectly executed by an intelligent murderer. Instead, it begins as a fight about trivial matters of honor, status, and reputation between men (such as when one man insults another or makes moves on another's girlfriend). Fights escalate because neither is willing to back down, until they become violent and one of the men ends up dead. Because women prefer to mate with men of high status and good reputation,[9] a man's status and reputation directly correlates with his reproductive success. Men are therefore highly motivated (albeit unconsciously) to protect their honor, and often go to extreme lengths to do so. Daly and Wilson thus explain homicides between men in terms of their (largely unconscious) desire to protect their status and reputation in their attempt to gain reproductive access to women.

Incidentally, this is why many evolutionary psychologists believe that the death penalty does not deter murder. The logic of the death penalty assumes that most murders are premeditated. A potential murderer carefully and rationally weighs the costs and benefits of the act, and decides not to murder if the costs outweigh the benefits. This might describe a fictional murderer on *Columbo*, but not real-life murderers, who do not stop to think before escalating their trivial altercations into fatal fights.

The logic of the death penalty also assumes that execution is the worst fate possible. From the evolutionary psychological perspective, there is something worse than death, which is the total reproductive failure that awaits any man who does not compete for mates in a polygynous society. If they compete and fight with other men, they *may* die (by being either killed by the other man or executed by the state); if they don't compete, however, they *will definitely* die, reproductively, by leaving no copies of their genes.

Rape may appear to be an exception to this reasoning because, unlike murders and assaults, the victims of rape are usually women, and there is therefore no male competition for status and reputation. However, the same psychological mechanism that compels men to gain reproductive access to women by competing with each other can also motivate men to commit rape. Predatory rapists are overwhelmingly men of lower class and status, who have very dim prospects of gaining legitimate reproductive access to women.[10] While it is not a manifestation of competition and violence, rape may be motivated by men's psychological mechanisms that urge them to gain reproductive access to women when they do not have the legitimate means to do so.

We can also extend the same analysis to property crimes. If women prefer to mate with men with more resources, then men can increase their reproductive prospects by acquiring material resources. Resources in traditional societies, however, tend to be concentrated in the hands of older men; younger men are often excluded from attaining them through legitimate means and must therefore resort to illegitimate means. One method of doing so is to appropriate someone else's resources by stealing them. So the same psychological mechanism that motivates violent crime can also motivate property crimes.

### *Crimes Evolved Before Norms Against Them*

Our suggestion that men steal in order to attract women might at first glance appear strange, since theft and other forms of resource extortion are universally condemned in human societies; in fact, such condemnation is another cultural universal. It is quite possible, however, that the psychological mechanism that motivates young males to commit violent and property crimes evolved in our ancestors in evolutionary history before the ape-human split (five to eight million years ago), even before the ape-monkey split (fifteen to twenty million years ago). In fact, our reasoning logically requires that the crucial psychological mechanisms emerge *before* the informal norms against violence and theft do; otherwise, violent competition and accumulation of resources through theft would not lead to higher status and reproductive success for males because they would be ostracized for violating the norms. We believe that the norms against violence and theft might have evolved *in reaction to* the psychological mechanisms that compelled young males to engage in violence and theft. The fact that violent and predatory acts that humans would classify as criminal are quite common among nonhuman species that do not have informal norms against such acts increases our confidence in this suggestion.[11]

These are some of the reasons why men are more violent and criminal. Crime and violence pay in reproductive terms, by allowing men to eliminate or intimidate their rivals and to accumulate resources to attract mates when they lack legitimate means to acquire such resources. But that is just one side of the coin. What about women? Given what we present above, why would *any* woman commit crimes at all?

### *Staying Alive*

The evolutionary psychologist Anne Campbell offers the "staying alive" theory of female criminality, which answers these questions and more.[12] Her theory begins with the fundamental observation that offspring's survival and thus reproductive success depends more heavily on maternal than paternal care and investment. It is therefore imperative for mothers rather than fathers to survive long enough to take physical care of their offspring to ensure their survival to sexual maturity. This, Campbell argues, is why females are more risk-averse than males are. The potential benefit of taking risks–by engaging in physical competition for resources and mates, for instance–simply does not justify the potential cost (the very survival of their offspring, which heavily hinges on the mother's own survival). A woman's primary goal is therefore to stay alive for the sake of her children.

Campbell goes on to point out, however, that females do occasionally need to compete for resources and mates, especially when these are scarce. This is why women sometimes compete for "a few good men," and occasionally resort to violence and theft to achieve their goals, even though, consistent with their primary goal to stay alive, their tactics of competition are usually low-risk (larceny rather than robbery) and indirect (spreading negative gossip and rumors about a rival behind her back rather than direct physical confrontation with her).

In her most recent work, Campbell[13] goes even further toward theoretical integration of male and female criminality. She argues that men and women do not differ in the *benefits* of aggression: high-status men who are winners of male competition may get access to mates and thus more opportunities for sex, but high-status women who are winners of female competition may get priority access to resources and greater protection afforded by high-status males. In other words, Campbell argues, women must compete for high-quality mates just as much as men do. It is therefore only the *costs* of aggression, Campbell argues, that distinguish men and women, and explain the far lower incidence of aggression among women.

Campbell points out that "theft by women is usually tied to economic need and occurs as part of their domestic responsibilities for their children," whereas "robbery is the quintessential male crime, in which violence is used both to extract resources and to gain status."[14] Apart from their tendency and inclination to avoid physical risks and danger altogether, this is another reason that women commit fewer crimes than men. Women only steal what they need for them and their children to survive, whereas men steal to show off and gain status as well as resources. In other words, *women steal less than men for exactly the same reason as they earn less than men.*[15] (See "Why Do Men Earn More Money and Attain Higher Status Than Women?" in chapter 7.) Women generally earn less than men do because they tend to make only what they need and usually have better things to do than earn money, whereas men are motivated to earn far more than they need to survive in order to use it to attract women. Similarly, women steal less than men do because they tend to steal what they need to survive and do not use crime for other purposes, like showing off and gaining status.

The work of evolutionary psychologists Martin Daly, Margo Wilson, and Anne Campbell thus explains why men are so much more violent and criminal than women are, and why this sex difference is culturally universal. We should point out, however, that according to the Interpol data, there is one exception to this rule in the world. A significant minority or even majority of offenders of *all* serious felonies in Syria *year after year* are women. We are frankly baffled by these statistics; however, it is very difficult for us (or any evolutionary psychologist) to believe that Syrian women, alone in the whole world, are genuinely more criminal than women elsewhere. We strongly suspect that either these statistics reflect some clerical error (for example, "male" and "female" were wrongly labeled when the Interpol form was first translated into Arabic many years ago, and the same mislabeled forms are photocopied and used every year) or there are some cultural or institutional reasons (for example, women may routinely take the fall for crimes committed by their husbands, brothers, or fathers). We have asked several Syrian experts for a possible explanation since we first noticed this statistical anomaly nearly a decade ago. We have not found one; however, we suspect that Syrian women do *not* commit the majority of serious crimes in their country.

## Q. What Do Bill Gates and Paul McCartney Have in Common with Criminals?

For nearly a quarter of a century, criminologists have known about a persistent empirical phenomenon called the "age-crime curve." In their highly influential 1983 article "Age and Explanation of Crime," two leading criminologists, Travis Hirschi and Michael R. Gottfredson, claim that the relationship between age and crime is the same across all social and cultural conditions at all times. In every society, for all social groups, for all races and both sexes, at all historical times, the tendency to commit crime and other analogous, risk-taking behavior rapidly increases in early adolescence, peaks in late adolescence and early adulthood, rapidly decreases throughout the 20s and 30s, and levels off during middle age. Although there have been minor variations observed around the "invariant" age-crime curve,[16] the essential shape of the curve for serious interpersonal crimes is widely accepted by criminologists.[17]

### *Everyone Wants a Piece of the Action*

One of the striking features of the age-crime curve is that it is not limited to crime. The same age profile characterizes "every *quantifiable* human behavior that is *public* (i.e., perceived by many potential mates) and *costly* (i.e., not affordable by all sexual competitors)."[18] The relationship between age and productivity among *male* jazz musicians, *male* painters, *male* writers, and *male* scientists, which might be called the "age-genius curve,"[19] is essentially the same as the age-crime curve.[20] Their productivity–the expressions of their genius–quickly peaks in early adulthood, and then just as quickly declines throughout adulthood. The age-genius curve among their female counterparts is much less pronounced and flatter; it does not peak or vary as much as a function of age.

It is not difficult to find personifications of the age-genius curve. Paul McCartney has not written a hit song in years, and now spends much of his time painting. Bill Gates is now a respectable businessman and philanthropist, and is no longer the computer whiz kid of his earlier years. J. D. Salinger now lives as a total recluse and has not published anything in more than three decades. Orson Welles was mere 26 when he wrote, produced, directed, and starred in *Citizen Kane*, which many consider to be the greatest movie ever made. There are some exceptions. Many artists, writers, and scientists remain productive into their middle and old ages, just like there are a few career criminals who commit crimes all their lives. But, in general, the pattern of youthful productivity holds for most.

What is the reason behind all this? Why do criminals usually desist from committing crimes as they age? Why does the productivity of creative geniuses also often fade with age? A single evolutionary psychological theory can explain the productivity of both creative geniuses and criminals over the life course.[21] According to this theory, *both crime and genius are expressions of young men's competitive desires, whose ultimate function in the ancestral environment would have been to increase reproductive success.*

### *What Explains the Crime and Genius Curves?*

As we've discussed, there are reproductive benefits of intense competitiveness to men. In the physical competition for mates, those who are competitive may act violently toward their male rivals. Their violence serves the dual function of protecting their status and honor, and discouraging or altogether eliminating their rivals from future competition for mates. Their competitiveness also inclines them to accumulate resources to attract mates by stealing from others, and the same psychological mechanism can probably induce men who cannot gain legitimate access to women to do so illegitimately through forcible rape. Men who are less inclined toward crime and violence may express their competitiveness through their creative activities in order to attract mates.[22]

There are no reproductive benefits from competition before puberty because prepubescent males are not able to translate their competitive edge into reproductive success. With puberty, however, the benefits of competition rapidly increase. Once the men are reproductively capable, every act of competition (be it through violence, theft, or creative genius) can potentially augment their reproductive success. The benefits of competition stay high after puberty for the remainder of their lives because human males are reproductively capable for most of their adult lives.

### *The Downside of the Curve*

This is not the whole story, however. There are also costs associated with competition. Acts of violence can easily result in the man's own death or injury, and acts of resource appropriation can trigger retaliation from the rightful owners of the resources. A man's reproductive success is obviously compromised if the competitive acts result in his death or even injury. Before men start reproducing (before their first child), there are few costs of competition. True, being competitive might result in their death or injury, and they might therefore lose in the reproductive game if they are too competitive. However, they also lose by *not* competing. If they do not compete for mates in a polygynous society, which all human societies are (see "Why [and How] Are Contemporary Westerners *Polygynous*?" in chapter 4), they will be left out of the game and end up losing as a result. In other words, young men *might* lose if they are competitive, but given polygyny, they will *definitely* lose if they are not. So there is little cost to being competitive, even at the risk of death and injury; the alternative is worse in reproductive terms, which once again is the reason the death penalty cannot deter young men.

The cost of competition, however, rises dramatically with the birth of the first child and subsequent children. True, men still benefit from competition because such acts of competition might attract additional mates even after their initial reproduction. However, a man's energies and resources are put to better use by protecting and investing in his existing children. In other words, with the birth of children, men should shift their reproductive effort away from mating and toward parenting. If the men die or get injured in their acts of competition, their existing children will suffer; they might starve without their father's parental investment or fall victim to predation by others without their father's protection. The costs of competition therefore rapidly increase after the birth of the first child, which usually happens several years after puberty because men need some time to accumulate enough resources and attain sufficient status to attract their first mate. Nevertheless, in the absence of artificial contraception, reproduction probably began at a much earlier age in the ancestral environment than it does today. There is therefore a gap of several years between the rapid rise in the benefits of competition and the similarly rapid rise in its costs.

Both the age-crime curve and the age-genius curve can be explained as the mathematical difference between the benefits and costs of competition. Young men rapidly become more violent, more criminal, and creatively more expressive in late adolescence and early adulthood as the benefits of competition rise, but then their productivity just as rapidly declines in late adulthood as the costs of competition rise and cancel its benefits. Criminality, genius, and productivity in virtually everything else men do vary as they do over the life course because they represent the difference between the benefits and costs of competition.

These calculations have been performed by natural and sexual selection, so to speak, which then equips male brains with a psychological mechanism to incline them to be increasingly competitive immediately after puberty and to make them less competitive right after the birth of the first child. Men simply do not *feel like* acting violently, stealing, or conducting additional scientific experiments, or they just *want to* settle down after the birth of the child, but they do not exactly know why. The intriguing suggestion here is that a single psychological mechanism may be responsible for much of what men do, whether they are criminals or scientists.[23]

### We All Have the Winners' Genes

Now, given that human society has always been mildly polygynous, there were many men who did not succeed at securing mates and reproducing. These men had everything to gain and nothing to lose by remaining competitive and violent for their entire lives. However, *we are not descended from these men.* By definition, we are all exclusively descended from men (and women) who attained some reproductive success—none of us are descended from total reproductive losers who left no offspring—and we are *disproportionately* descended from those who attained great reproductive success. (Twelve children carry the genes of a man who had twelve children, but only one child carries the genes of a man who had only one child. And, of course, no children carry the genes of a man who had no children. Yes, childlessness is perfectly heritable!) Contemporary men, therefore, did not inherit from reproductive losers psychological mechanisms that force them to stay competitive and keep trying to secure mates for their entire lives.

## *Female Choice*

The similarity between Bill Gates, Paul McCartney, and the criminals (in fact, *all* men) in evolutionary history points to a very important concept in evolutionary biology: female choice. In all species in which the female makes greater parental investment than the male (such as humans and all other mammals), mating is a female choice; it happens when the female wants it to happen, and with whom she wants it to happen, not when the male wants it to happen.[24]

The power of female choice becomes quite apparent in a simple thought experiment. Imagine for a moment a society where sex and mating were entirely a male choice; individuals have sex whenever and with whomever men want, not whenever and with whomever women want. What would happen in such a society? *Absolutely nothing*, because people would never stop having sex! There would be no civilization in such a society, because people would not do anything besides have sex. This, incidentally, is the evolutionary explanation for why gay men tend to have significantly more sex partners and have sex significantly more frequently than straight men do: because there are no women in their relationship to say no.[25] Sexually active straight men on average have had 16.5 sex partners since age 18; gay men have had 42.8.

In reality, however, women do often say no to men. This is why men throughout history have had to conquer foreign lands, win battles and wars, compose symphonies, author books, write sonnets, paint portraits and cathedral ceilings, make scientific discoveries, play in rock bands, and write new computer software in order to impress women so that they will agree to have sex with them.[26] There would be no civilizations, no art, no literature, no music, no Beatles, no Microsoft, if sex and mating were a male choice. Men have built (and destroyed) civilizations in order to impress women so that they might say yes. Women are the reason men do everything.

The comedian Bill Maher captures the essence of female choice perfectly, when he quips: "For a man to walk into a bar and have his choice of any woman he wants, he would have to be the ruler of the world. For a woman to have the same power over men, she'd have to do her hair." In other words, any reasonably attractive young woman exercises as much power as does the (male) ruler of the world.

## Q. Why Does Marriage "Settle" Men Down?

There is something else that crime and genius have in common. (See "What Do Bill Gates and Paul McCartney Have in Common with Criminals?" above.) Just as age does, *marriage depresses both tendencies.*

Criminologists have long known that criminals tend to "settle down" and desist (stop committing crime) once they get married, while unmarried criminals continue their criminal careers. But criminologists tend to explain this phenomenon from the social control perspective[27] pioneered by the criminologist Travis Hirschi[28] (the same Hirschi of the team who first discovered the age-crime curve). Social control theorists argue that marriage creates a bond to the conventional society, and investment in this bond, in the form of a strong marriage, makes it less likely that the criminal would want to remain in the criminal career, which is incompatible with the conventional life. Men must therefore desist from crime when they get married in order to protect their investment in conventional life; in Hirschi's language, married men develop a "stake in conformity."[29] Marriage also increases the scope and efficiency of social control on the criminal. Now there is someone living in the same house and monitoring the criminal's behavior at all times. It would be more difficult for the criminal to escape the wife's watchful eye and engage in illicit activities.

The social control explanation for the effect of marriage on desistance from crime makes perfect sense, *until* one realizes that marriage has the same desistance effect on perfectly legal, conventional activities, such as science. A comparison of the "age-genius curve" among scientists who were married at some point in their lives with the same curve among those who never married shows the strong desistance effect of marriage on scientific productivity. Half as many (50.0 percent) unmarried scientists make their greatest contributions in their late 50s as they do in their late 20s. The corresponding percentage among the married scientists is 4.2 percent. The mean age of peak productivity among the unmarried scientists (39.9) is significantly later than the mean peak age among married scientists (33.9).[30]

Given that the Nobel Prize for scientific achievement didn't exist in the ancestral environment, the evolved psychological mechanisms of men appear to be rather precisely tuned to marriage as a cue to desistance. Nearly a quarter (23.4 percent) of all married scientists make their greatest scientific contribution in their career and then desist within five years after their marriage. The mean *delay* (the difference between their marriage and their peak productivity) is a mere 2.6 years; the median is 3.0 years. It therefore appears that scientists rather quickly desist after marriage, while unmarried scientists continue to do important scientific work. When you remember that great scientific discoveries usually require many years of cumulative and continued research, the near coincidence of the male scientists' marriages and their desistance (after which they cease to make any greater scientific discoveries) is remarkable. A study by the sociologists Lowell L. Hargens, James C. McCann, and Barbara F. Reskin also demonstrates that childless research chemists are more productive than those with children.[31]

You may think that unmarried scientists continue to make scientific contributions much later in their lives because they have more time to devote to their careers. Unmarried and therefore childless scientists do not have to spend time taking care of their children, driving them back and forth between soccer practices and ballet lessons, or doing half the household chores, and that is why unmarried scientists can continue making great contributions to science while married scientists must desist to devote their time to their families. This is precisely Hargens et al.'s interpretation of the negative correlation between parenthood and productivity among research chemists.

But we would point out that almost all the scientists in the main data on scientific biographies we rely on above lived in the eighteenth and nineteenth centuries, when married men made very little contribution in the domestic sphere and their wives did not have their own careers. Hargens et al.'s data come from 1969–70, when this was probably still true to a large extent. We would therefore suggest that, if anything, married scientists probably had *more* (rather than less) time to devote to science, because they had someone to take care of their domestic needs at all times.

Why, then, does marriage depress the productivity of all men, criminals, and scientists alike? What underlies the desistance effect of marriage?

### *Enjoying the Fruits of Their Labor*

The social control perspective on the desistance effect of marriage is at best incomplete if marriage has the same desistance effect on scientists. Unlike criminal behavior, scientific activities are completely within conventional society and are thus not at all incompatible with marriage and other strong bonds to conventional society. Unlike criminals, scientists are not subject to social control (by their wives or others), since scientific activities are not illegal or deviant in any way.

We believe an evolutionary psychological theory provides a much simpler explanation for the desistance effect of marriage for both crime and science, in the form of a single psychological mechanism that predisposes young men to compete and excel early in their adulthood but subsequently turns off after the birth of their children (which quickly followed pair-bonding and regular sex in the absence of reliable means of birth control in the ancestral environment). After their marriage and children, male scientists do not *feel like* spending hours and hours in their labs, just like married criminals do not *feel like* taking great risks and committing crimes. But neither scientists nor criminals know why.

From the evolutionary psychological perspective, reproductive success is the end, and everything men do (be it crime or scientific research) is but a means to this ultimate end. From this perspective, the question of why marriage depresses crime and scientific productivity misses the whole point. Does it make sense for men to continue employing the means even after they have achieved the ends they were trying to attain with the means? This is why married men are less likely to engage in a whole range of risk-taking behaviors, like driving fast, which are designed indirectly and unconsciously to attract women. Indeed, automobile insurance statistics clearly show that married men have fewer car accidents.

## Q. Why Do Some Men Beat Up Their Wives and Girlfriends?

Critics of evolutionary psychology often claim that evolutionary psychological explanations are "untestable" and "unfalsifiable."[32] As but one perfect example of eminent testability and falsifiability of evolutionary psychological explanations[33] we offer two competing explanations of domestic violence, formulated by the two deans of modern evolutionary psychology (who happen to be married to each other, no less).

When Martin Daly and Margo Wilson began studying domestic violence and uxoricide (the killing of one's wife) in the early 1980s, they had competing explanations.[34] Daly hypothesized that domestic violence and uxoricide resulted when the husband did not value his wife sufficiently and mistreated her as a result. Since a wife's fertility and reproductive value decline with age, Daly predicted that older wives were at a greater risk of spousal abuse and homicide than younger wives. Wilson, in contrast, hypothesized that domestic violence and uxoricide were a maladaptive byproduct of the husband's inclination and tendency to guard his wife to make sure that she did not have sexual contact with other men. Because men should be more motivated to guard younger, more valuable wives, Wilson predicted that younger wives were at a greater risk of spousal abuse and homicide than older wives.

Both explanations use impeccable evolutionary psychological logic and derive from known facts, but both predictions could not be true simultaneously. So Daly and Wilson got to work as the good scientists that they are, collecting data on domestic violence and uxoricide in Canada and the United States, and putting the two competing predictions to the empirical test. Their data showed that younger wives were at a much greater risk of violence and murder than older wives. In the end, Wilson's prediction turns out to be true, and Daly's false.[35] Is evolutionary psychology untestable and unfalsifiable?

Astute readers may be thinking right now, "But younger women are usually married to younger men. And younger men are more violent than older men, as you point out in your discussion of the age-crime curve (see "What Do Bill Gates and Paul McCartney Have in Common with Criminals?" above). So younger women are at a greater risk of spousal abuse and murder, not because they are young but because their husbands are young and therefore more violent."

Close, but no cigar. While it is difficult to separate the effects of the husband's age and the wife's age, careful statistical analyses show that the wife's age almost entirely determines the likelihood of being a victim of spousal abuse and homicide. Middle-aged husbands (ages 45–54) legally married in Canada to much younger wives (ages 15–24) are *more than six times* as likely to kill their wives than young husbands (ages 15–24) married to women of similar age.[36] Among common-law marriages, middle-aged husbands married to much younger wives are *more than 45 times* as likely to kill their wives as young husbands.[37] *The effect of the wife's age is so powerful that it overrides and even reverses a man's tendency to become less violent with age.* Thus, while it is true that younger men in general are much more violent and commit more murders than older men, young and old men kill different types of people. Young men kill other men (their male sexual rivals); older men kill their wives. As a result, the proportion of men among murder victims declines as the murderer's age increases. For murderers aged 15–19, 86.3 percent of the victims are males; for murderers aged 65–69, only 51.4 percent of them are males.[38]

## *An Adaptation Gone Awry*

From the evolutionary psychological perspective, spousal abuse is an extreme, maladaptive, and largely unintended consequence of a man's desire for mate-guarding. Because of the possibility of cuckoldry (unwittingly investing in someone else's genetic offspring), men are strongly motivated to guard their mates to make sure that they do not have sexual access to other men. And they use any means, including intimidation and violence, to achieve this goal.[39] Unfortunately, sometimes their adaptive strategy of mate-guarding goes too far and results in a maladaptive outcome of spousal abuse and even murder. Because young women are reproductively more valuable than older women, men are more motivated to protect and guard their younger wives than their older wives, with the unfortunate consequence that younger wives are at a greater risk of spousal abuse than older wives. This is why it is the wife's age, not the husband's, that predicts the likelihood of spousal abuse and murder. Even though a 50-year-old man is typically much less violent and criminal than a 25-year-old man, a 50-year-old man married to a

25-year-old woman is much more likely to abuse and murder his wife than a 25-year-old man married to a 50-year-old woman (although there are very few such couples) or even a 25-year-old man married to a 25-year-old woman.[40]

## *The Myth of the Midlife Crisis*

This is an excellent opportunity for us to shed evolutionary psychological light on a common misunderstanding, since it allows us to shift our attention from a dark topic like domestic violence and apply the same logic to a much lighter topic: the midlife crisis. Many believe that men go through a midlife crisis when they are in midlife (in middle age). Not quite. Many middle-aged men *do* go through midlife crisis, but it's not because *they* are middle-aged but because their wives are. Just as it is the wife's age, not the husband's, that determines the risk of spousal abuse and murder, it is the wife's age, not the husband's, that prompts the constellation of behavior commonly known as a "midlife crisis." From the evolutionary psychological perspective, a man's midlife crisis is precipitated by his wife's imminent menopause and the end of her reproductive career, and thus his renewed need to attract younger, reproductive women. Accordingly, a 50-year-old man married to a 25-year-old women would not go through a midlife crisis, while a 25-year-old man married to a 50-year-old woman would, just like a more typical 50-year-old man married to a 50-year-old woman would. It is not his midlife that matters; it is hers. So when he buys a shiny red sportscar, he's not trying to regain his youth; he's trying to attract young women to replace his menopausal wife by trumpeting his flash and cash.

# 7

# Life's Not Fair, or Politically Correct

## THE EVOLUTIONARY PSYCHOLOGY OF POLITICAL AND ECONOMIC INEQUALITIES

The topics covered in the final two chapters of this book (chapters 7 and 8) are the least explored areas of application for evolutionary psychology. Because of its origin in the field of psychology and its emphasis on sex and mating, most of the scientific progress and discoveries in evolutionary psychology have been on individual behavior and cognitions–how men and women behave differently, how the human brain perceives the world, the biases and tendencies in our thinking, and so on. Most of the applications of evolutionary psychology in the social sciences have therefore been "micro"–on the small scale of individuals.

There have not been many "macro" applications of evolutionary psychology–to the issues of economy, politics, and society at large. However, there have been some very intriguing studies in this area as well. Because both of us were originally sociologists, concerned more with macro issues than micro issues, this is where our training in sociology meets our current interest in evolutionary psychology.

One fascinating discovery from the application of evolutionary psychology to macro issues is that what we often regard as "beyond" individuals–because they are so much bigger than them–such as issues related to social institutions, economic and political inequalities, social problems, wars, religion, and even culture itself, have the same origins as individual behavior and cognitions. They all stem from our evolved psychological mechanisms in our brains. They are all macro manifestations of our human nature and biology.

## Q. Why Do Politicians Risk Everything by Having an Affair (but Only If They Are Male)?

On the morning of Wednesday, January 21, 1998, Americans woke up to breaking news. The *Washington Post*, one of the nation's leading newspapers, reported the allegation that President Bill Clinton had an affair with a 24-year-old White House intern. On that January morning, as the story unfolded in front of the stunned nation, America and the rest of the world had not yet had an inkling of what was in store: a yearlong political scandal that consumed the nation (and the world) and culminated on December 19, with Clinton being impeached by the House of Representatives–the first elected President ever to be impeached in American history.[1]

While the whole nation was in shock, one woman in Michigan woke up to the news on the morning of January 21, 1998, sipped her coffee while watching the events unfold on TV, smiled to herself, and said, "I told you so." She is the Darwinian historian Laura L. Betzig. For more than twenty years, Betzig has written on the mating behavior and reproductive success of politicians and other political leaders in history.[2] She points out that while powerful men throughout Western history have *married* monogamously (they had only one legal wife at a time), they have always *mated* polygynously (they had lovers, concubines, and female slaves).[3] Many had harems, consisting of hundreds and even thousands of virgins. With their wives they produced legitimate heirs; with the others they produced bastards (Betzig's term). Genes and inclusive fitness make no distinction between the two categories of children. While the legitimate heirs, unlike the bastards, inherited their fathers' power and status and often went on to have their own harems, powerful men sometimes invested in their bastards as well.

As a result, powerful men of high status throughout human history attained very high reproductive success, leaving a large number of offspring (legitimate or otherwise), while countless poor men in the countryside died mateless and childless. Moulay Ismail the Bloodthirsty, whom we encountered in chapter 2, stands out *quantitatively*, having left more offspring than anyone else on record, but he was by no means *qualitatively* different from other powerful men, like Bill Clinton.

### *Why Not?*

From Betzig's Darwinian historical perspective, the question that many Americans and others throughout the world asked in 1998, "Why on earth would the most powerful man in the world jeopardize his job for an affair with a young woman?" is a silly question. Betzig's answer would be: Why not?

Recall from chapter 1 ("What Is Evolutionary Psychology?") that the underlying motive of all human behavior is reproductive; reproductive success is the purpose of all biological existence, including humans.[4] Humans do much of what they do, directly or indirectly, knowingly or (usually) unknowingly, to achieve reproductive success.

Attaining political office is no exception. From this perspective, men strive to attain political power (as Bill Clinton did all his life, since his fateful encounter with John F. Kennedy at the White House in 1963), consciously or unconsciously, in order to have reproductive access to a larger number of women. In other words, reproductive access to women is the *goal*, political office is but one *means*. To ask why the President of the United States would have a sexual encounter with a young woman is like asking why someone who worked very hard to earn a large sum of money would then spend it. The purpose of earning money is to spend it. The purpose of becoming the President (or anything else men do) is to have a larger number of women with whom to mate.

What distinguishes Bill Clinton is not that he had extramarital affairs while in office; others have, and more will in the future. It would be a Darwinian puzzle if they did not. What distinguishes Clinton instead is that he got caught and that his affair became a spectacular political scandal. What Clinton's genes did not know is that he was not permitted by others to have sex with a large number of women and that he could not get away with it when most of his predecessors have, like all the kings, emperors, sultans, and democratically elected presidents whose reproductive lives Betzig's work describes in great detail. Clinton's genes didn't know about the DNA fingerprinting technology that ultimately exposed the affair and forced him to admit it publicly, because no such thing existed in the ancestral environment.

## Q. Why Do Men So Often Earn More Money and Attain Higher Status Than Women?

In all industrialized nations, women earn less money and attain lower occupational status than men do.[5] This is true across the board, among blue-collar and white-collar workers and professionals, and in capitalist, socialist, and communist economies. Why?

### *The Traditional Social Science View*

The sex difference in earnings is one of the central concerns of economics[6] and sociology.[7] Economists and sociologists identify three different parts to the total difference in earnings between men and women. First, there is the difference in what they call "human capital"–education, job skills, training, and other individual traits that affect productivity and job performance. Second, sex difference in earnings can be due to occupational segregation by sex–the fact that men and women tend to occupy different jobs. Men tend to occupy "blue-collar" jobs (manufacturing, construction, truck driving), while women tend to occupy "pink-collar" jobs (secretarial, nursing, teaching). Third, the sex difference in earnings can be due to sex discrimination, where employers pay equally qualified men and women doing the same job differently.

To the extent that the sex gap in pay is due to differences in human capital and productivity, it is considered to be fair by most social scientists. To the extent that the sex gap in pay results from the existence of blue- and pink-collar jobs, then paying all workers in a given occupation equally will not close the total sex difference in earnings. Paying the same wages to male and female truck drivers and to male and female secretaries will not close the sex gap in pay if truck drivers make more than secretaries and most truck drivers are male and most secretaries are female. The existence of occupational sex segregation thus requires consideration of "comparable worth."[8]

Because they are deeply wedded to the Standard Social Science Model, most economists and sociologists assume that men and women are on the whole identical in their preferences, values, and desires. They therefore assume that any remaining sex difference in earnings that is not due to sex differences in human capital or sex segregation on the job must be due to employer discrimination. The existence of discrimination, however, must always be inferred from statistical evidence and cannot be observed directly. Social scientists are not likely to witness an employer telling the employees, "I'm paying you more because you are a man, and I'm paying you less because you are a woman." Nor are employers likely to admit to such a practice if they indeed engaged in it.

### *But Men and Women* Are *Different*

The conclusion that there is sex discrimination by employers crucially depends on the assumption that men and women are on average identical, except in their amount of human capital (education, job experience, skills) and the jobs they hold. If, on the other hand, men and women with the same amount of human capital and in the same jobs are nonetheless *inherently* and *fundamentally* different in ways that affect their earnings, for instance in their preference and desire for earning money, then discrimination becomes unnecessary to explain the sex gaps in pay. If men and women are different in *internal* preferences and dispositions, such as their desire and drive to earn money, then no *external* factors, such as employer discrimination or a "glass ceiling," becomes necessary to explain the sex difference in earnings.

The legal scholar Kingsley R. Browne has pioneered evolutionary psychological work on the sex differences in the workplace, such as earnings and occupational sex segregation.[9] (We encounter his work on sexual harassment later in this chapter, "Why Is Sexual Harassment So Persistent?") Browne points out that because of differential selective pressures that men and women faced throughout evolutionary history, men and women have evolved to possess different temperaments. Throughout evolutionary history, material resources and higher status were a man's essential means to reproductive success, because women preferred to mate with resourceful men of high status who could protect and invest heavily in their children. In contrast, physically taking care of children was a woman's means. As a result, women today, who inherited their psychological mechanisms from their female ancestors, are far less risk-taking (because if their ancestors engaged in risky behavior and got injured or killed as a result, their children most likely died),[10] less status-seeking (because status did not enhance women's reproductive success), and less aggressive and competitive (because throughout evolutionary history, men competed to gain access to women, not the other way around).

Browne suggests that men are much more single-mindedly devoted to earning money and achieving higher status than women are. In a study of an American sample, men are significantly more likely to rank income as an important criterion for selecting a job than women are. The absolute sex difference is greater among teenagers than among older workers, so it is not a realistic response to a lifetime's experience of earning less than men, as feminists and other conventional social scientists might contend.[11] In contrast, women place significantly greater emphasis on the criterion "the work is important and gives me a feeling of accomplishment" for selecting a job.[12] As Anne Moir and David Jessel, authors of *Brain Sex: The Real Difference Between Men and Women*, state: "In the end, the secret of male achievement in the world of work probably lies in the relative male insensitivity to the world of everything–and everybody–else."[13]

Browne reminds us that many jobs that pay higher wages require their occupants to work longer hours, relocate to new cities without regard to consequences for family and children (for white-collar or professional jobs), or work in dangerous and unpleasant conditions (for blue-collar workers). It is not that women do not want money or prefer less money to more; nobody in their right mind does. It is instead that women are unwilling to pay the price and make the necessary sacrifices (often in the welfare and well-being of their children) in order to advance in the corporate hierarchy and earn more money. Once again, Moir and Jessel put it best: "Men who fail will often offer the excuse that 'Success isn't worth the effort.' To the female mind, this is not so much an excuse as a self-evident truth."[14] In other words, men make more money because they want to; women make less money because they have better things to do than make money.[15]

The sex gap in earnings and the so-called glass ceiling are caused not by employer discrimination or any other external factors, but by the sex differences in internal preferences, values, desires, dispositions, and temperaments. Just as there are a few exceptional women who are more single-mindedly motivated to earn money and attain higher status than the average man, so too are there a few women who make more money and attain higher status than most men.

From the 1960s through the 1980s, feminists claimed that women earned only 59 cents for every dollar earned by men.[16] The precise figure has since been revised upward to 64 cents in 1986,[17] 70 cents in 1987,[18] and, according to President Clinton (if he counts as a feminist), 75 cents in 1999,[19] but their claim is that women still earn substantially less than men do. However, all of these comparisons ignore the inherent sex differences in dispositions and temperaments. More careful statistical comparisons of men and women who are equally motivated to earn money show that women now earn 98 cents for every dollar men make,[20] and sex has no statistically significant effect on workers' earnings.[21] Adjusted for occupation and motivation, men today do not earn significantly more than women do.

Just as most women are not as single-mindedly motivated to earn money and attain higher status than the average man, most women do not earn as much money and attain as high status as men do. Browne rhetorically asks the question, "Once one breaks the glass ceiling, does it still exist?"[22] In liberal capitalist societies like the US and the UK, both men and women are free to pursue what they want. They just tend to want to achieve different things.

## Q. Why Are Most Neurosurgeons Male and Most Kindergarten Teachers Female?

In a series of scientific articles and books, and in the popular science book *The Essential Difference*, the Cambridge psychologist and autism researcher Simon Baron-Cohen advances the "extreme male brain" theory of autism.[23] His theory can simultaneously account for many (though not all) clinical manifestations of autism (such as exhibiting severe deficits in interpersonal domains, while maintaining normal or even exceptional abilities in others) as well as the fact that an overwhelming majority of autistics are male.

Baron-Cohen's theory begins with the two crucial concepts of the *male brain* and the *female brain*. The male brain is primarily designed for *systemizing*, and the female brain is primarily designed for *empathizing*. What are systemizing and empathizing?

"Systemizing" is the drive to analyze, explore, and construct a system. The systemizer intuitively figures out how things work, or extracts the underlying rules that govern the behavior of a system. The purpose of this is to understand and predict the system, or to invent a new one.[24] Baron-Cohen enumerates six different types of systems: technical systems (artifacts, machines); natural systems (ecology, geography); abstract systems (logic, mathematics); social systems (law, economics); organizable systems (classifications, taxonomies); and motoric systems (physical movements, such as playing musical instruments or throwing darts). His definition of what constitutes a system is therefore very comprehensive, and seems to include everything that has to do with things rather than people. By a "system," Baron-Cohen means anything that is governed by logical and systematic rules.[25]

In contrast, "empathizing" is the drive to identify another person's emotions and thoughts, and to respond to them with an appropriate emotion. Empathizing occurs when we feel an appropriate emotional reaction in response to the other person's emotions. The purpose of this is to understand another person, to predict his or her behavior, and to connect or resonate with him or her emotionally.[26] In other words, empathizing is about spontaneously and naturally tuning in to the other person's thoughts and feelings. A good empathizer can immediately sense when an emotional change has occurred in someone, what the causes of this might be, and what might make this particular person feel better or worse. A good empathizer responds intuitively to a change in another person's mood with concern, appreciation, understanding, comforting, or whatever the appropriate emotion might be. A natural empathizer not only notices others' feelings but also continually thinks about what the other person might be feeling, thinking, or intending.[27] Empathy is a defining feature of human relationships and also makes real communication possible.[28]

Having defined what systemizing and empathizing are, Baron-Cohen then describes the distribution of systemizing and empathizing skills among men and women. Both systemizing and empathizing skills are distributed normally among the general populations of men and women. Men have a higher mean of systemizing skills than women, while women have a higher mean of empathizing skills than men. However, the sex distributions of systemizing and empathizing skills substantially overlap. This means that while men on average are better at systemizing and women on average are better at empathizing, there are many men who are better empathizers than women and many women who are better systemizers than men.[29]

The distribution of height provides a perfect analogy. Height is distributed normally both among men and among women. Men have a higher mean height than women. However, the sex distributions of height overlap sufficiently so that, while most men are taller than the average women, there are some men who are shorter than the average woman and some women who are taller than the average man. In Baron-Cohen's theory, the systemizing and empathizing skills have similar distributions and sex differences.

### *Two Types of Brains*

Baron-Cohen defines the brain of someone who is better at systemizing than empathizing as the "type S" brain, or the male brain (even though not everyone who possesses the male brain is male), and the brain of someone who is better at empathizing than systemizing as the "type E" brain, or the female brain (even though not everyone who possesses the female brain is female). Baron-Cohen then suggests that the type S brain was particularly adaptive for ancestral men, because systemizing ability was necessary for inventing and making tools and weapons, and because low empathizing ability was helpful for tolerating solitude during long hunting and tracking trips, and for committing acts of interpersonal violence and aggression necessary for male competition.

Similarly, Baron-Cohen argues that the type E brain was adaptive for ancestral women, because empathizing ability facilitates various aspects of mothering, such as anticipating and understanding the needs of infants who could not yet talk, or making friends and allies in new environments, in which ancestral women found themselves upon marriage. (In the ancestral environment, women left their group and married into a neighboring group upon puberty, a practice necessary to avoid inbreeding.) Natural and sexual selection would therefore have favored ancestral men having the type S brain and ancestral women having the type E brain.

Baron-Cohen then explains autism (and other autism spectrum disorders such as Asperger's syndrome) as a result of possessing the "extreme male brain," which is exceedingly good at systemizing but correspondingly poor at empathizing. Not only does Baron-Cohen's conceptualization of autism as a manifestation of the extreme male brain explain many of the clinical features of autism, but it also explains why it is so much more prevalent among men than among women.

One of the discoveries by Baron-Cohen and his team of researchers is the high prevalence of physicists, engineers, and mathematicians among the families of autistics and those afflicted with Asperger's syndrome.[30] This is because brain types (type S vs. type E) are largely genetically heritable and therefore "run in the family," and because these professions require high systemizing skills. This seems to be why most scientists and engineers are men. Contrary to what the Standard Social Science Model's social scientists claim, it has very little to do with "gender socialization" and much more to do with sex-typical brain types. (Once again, "gender socialization"–to the extent that it is widely practiced–simply reinforces and solidifies the innate genetic differences between the male and female brains.) By the same token, any occupation that requires a large amount of empathizing skills, such as kindergarten, preschool, and elementary school teachers; nurses and other caretakers; and social workers, are more likely to be female, because the typical female (type E) brain is highly useful in such occupations.

In liberal capitalist societies like the US or the UK, both men and women try to pursue occupations and professions that best suit them. Because some men have type E brains and some women have type S brains, there are some male nurses and kindergarten teachers, and some female neuroscientists and engineers. However, a majority of those in "systemizing" occupations are men, and a majority of those in "empathizing" occupations are women, and Baron-Cohen's theory can explain why.

## Q. Why Is Sexual Harassment So Persistent?

One of the unfortunate consequences of the ever-growing number of women joining the labor force and working side by side with men is the increasing number of sexual harassment cases, particularly in the United States. Why is this? Is sexual harassment a necessary consequence of the sexual integration of the workplace? What is sexual harassment, anyway, and how can evolutionary psychology explain it?

As with the study of sex differences in earnings and the "glass ceiling" (see "Why Do Men Earn More Money and Attain Higher Status Than Women?" above), the evolutionary psychologist who pioneered the study of sexual harassment is Kingsley R. Browne.[31] Browne identifies two types of sexual harassment cases: the quid pro quo cases ("You must sleep with me if you want to keep your job or be promoted") and the "hostile environment" cases (where the workplace is deemed too sexualized for workers to feel safe and comfortable). While feminists and other Standard Social Science Model scholars tend to explain sexual harassment in terms of patriarchy and other nefarious ideologies,[32] Browne locates the ultimate cause of both types of sexual harassment in the sex differences in evolved psychological mechanisms and mating strategies, thereby "seeking roots in biology rather than ideology."[33]

Studies unequivocally demonstrate that men are far more interested in short-term casual sex than women. For example, in a classic study,[34] 75 percent of undergraduate men approached by an attractive female stranger agree to have sex with her; none of the women approached by an attractive male stranger do. Many men who would not go on a date with the stranger nonetheless agree to have sex with her. In another study,[35] men on average desire nearly twenty sex partners in their lifetimes; women desire less than five. Men on average seriously consider having sex with someone after only one week of acquaintance; women's average is six months.

The quid pro quo and similar types of harassment are manifestations of men's greater desire for short-term casual sex than women's, and their willingness to use any available means to achieve their goal. While feminists often claim that sexual harassment is "not about sex but about power,"[36] Browne astutely points out that it is both; it is about men using power to get sex. "To say that it is only about power makes no more sense than saying that bank robbery is only about guns, not about money."[37]

The male-female differences in the desire for short-term casual sex are exacerbated by another male-female difference in evolved psychological mechanisms: the woman's desire to understate her sexual desire in a particular man and to engage in "token resistance."[38] In one study,[39] nearly 40 percent of undergraduate women admitted to saying no to sexual advances from a man even though they actually wanted to have sex with him. More than a third of these cases where the women initially said no eventually resulted in consensual sex. As the late behavior geneticist Linda Mealey eloquently puts it: "That females are selected to be coy will mean that sometimes saying 'no' really does mean 'try a little harder.'"[40] Of course, women sometimes do mean no when they say no, but this isn't always the case.

### *Hostile Environment: When Men Are Equal-Opportunity Harassers*

Browne explains the incidence of sexual harassment cases of the second variety (hostile environment) as a result of the sex differences in what men and women perceive as "overly sexual" or "hostile." While the courts in the United States often employ the standard of a fictitious "reasonable person" to determine whether a given workplace constitutes a hostile environment, Browne points out that there is no such thing as a "reasonable person"; there is only a reasonable man and a reasonable woman. What a reasonable man and a reasonable woman perceive to be a hostile environment may be entirely different. Browne questions the exclusive focus on the alleged victim's perspective.

While many women legitimately complain that they have been subjected to abusive, intimidating, and degrading treatment by their male colleagues and employers, Browne points out that long before women entered the labor force, men subjected *each other* to such abusive, intimidating, and degrading treatment. Abuse, intimidation, and degradation are all part of men's unfortunate repertoire of tactics employed in competitive situations. In other words, men are *not* harassing women in this fashion because they are treating women differently from men (which is the definition of discrimination under which sexual harassment legally falls), but the exact opposite: men harass women precisely because they are *not* discriminating between men and women.

Because of all the media attention and the soaring costs of litigation, most American firms and universities now have sexual harassment policies that categorically prohibit any sexual relations between and among their employees. Browne makes a sharp observation in this connection. Although sexual harassment surveys typically ask whether the respondent has ever been subjected to unwanted sexual advances in the workplace, they seldom, if ever, ask whether she has been subjected to *welcome* sexual advances. The answer must commonly be in the affirmative, since a large number of workers find their romantic partners at work.[41]

Men's and women's behavior that sometimes results in charges of sexual harassment is most often simply part of the normal repertoire of human mating strategies. They work well most of the time (as when a large number of men and women find satisfactory long-term and short-term mates in their workplace) but occasionally result in miscommunication and misunderstanding due to the evolved differences between the sexes, which is then given the label of sexual harassment. (See "He Said, She Said: Why Do Men and Women Perceive the Same Situation Differently?" in chapter 3.) While it might deter some legitimately abusive behavior, the current sexual harassment policy commonly practiced in many American organizations, which categorically prohibits any sexual relations between employees, is therefore likely to be *detrimental to women's sexual interests as much as men's*, because such prohibition eliminates *welcome* sexual attention and advances along with the unwelcome.

# 8

# The Good, the Bad, and the Ugly

## THE EVOLUTIONARY PSYCHOLOGY OF RELIGION AND CONFLICT

The topics we explore in our last substantive chapter–those concerning religion and group conflict–are the least explored areas of application for evolutionary psychology, probably because these are most remote from the immediate concerns with sex and mating. The connection between sex and religion is not as obvious as the connection between sex and marriage, for example. However, there is a connection. There have been a small number of important studies in the area of religion and group conflict from the evolutionary psychological perspective, and they shed new and often surprising light, providing novel answers to old questions.

There have been two seemingly contradictory findings in the area of conflict between groups: racism is innate, but race is not. More accurately, *ethnocentrism* (the tendency to value one's own group and correspondingly to devalue other groups) is an evolved innate tendency that all humans have. If you have read this far in this book, you should know why by now. We are designed to promote our own reproductive fitness and spread our genes. There is no place for universal love of all people in cold Darwinian logic. So contrary to what social scientists and hippies alike proclaim, we *don't* learn to be a racist through parental socialization; we learn *not* to be one.

Even though the tendency to favor "ingroup" members at the cost of "outgroup" members is innate (although we can overcome it through socialization and conscious effort), what counts as "ingroup" and "outgroup" is not. In particular, a very ingenious experiment[1] has shown that we can erase racial categories that we normally use under the right circumstances. This makes perfect sense, in retrospect, when you remember that our ancestors evolved in a mostly racially homogeneous environment. Encountering people of different races on a daily basis is a very recent phenomenon in human evolutionary history, so there could not be innate categories for different races in our brain, as there are for age and sex.

We will get to the topic of group conflict shortly. But first we start with religion and where it came from in the first place. . . .

## Q. Where Does Religion Come From?

It may be tempting to believe that religion[2] is an adaptation (or, in our language, an evolved psychological mechanism) designed by evolution by natural and sexual selection, since there are genetic and biological bases of religion. All human societies practice religion (making it one of the cultural universals);[3] whether one is religious or not, especially in adulthood, is largely genetically determined;[4] and certain parts of the brain are involved in religious thoughts and experiences.[5] However, this explanation of religion as an adaptation runs into one significant problem: What is the adaptive problem that religion is designed to solve? Do religious people live longer or have greater reproductive success?[6] So far, no one has been able to point to an adaptive problem that religion is designed to solve.[7]

As a result, many recent evolutionary psychological theories on the origins of religious beliefs share the view that religion is not an adaptation in itself but a *byproduct* of other adaptations. In other words, these theories contend that religion itself did not evolve to solve an adaptive problem so that religious people can live longer and reproduce more successfully, but instead emerged as a byproduct of adaptations that evolved to solve unrelated adaptive problems.

These theories,[8] in part or in whole, go as follows: When our ancestors faced some ambiguous situation, such as rustling noises nearby at night or a large fruit falling from a tree branch and hitting them on the head, they could attribute them to impersonal, inanimate, unintentional forces (such as wind blowing gently to make the rustling noises among the bushes and leaves, the mature fruit falling by its own weight from the branch by the force of gravity and hitting them on the head purely by coincidence) or to personal, animate, intentional forces (a predator sneaking up on them to attack, an enemy hiding in the tree branches and throwing fruit at their head). The question is, which is it?

## *Two Different Ways to Get It Wrong*

Given that the situation is inherently ambiguous and could be caused by either intentional or unintentional forces, our ancestors could have made one of two possible errors. They could have attributed the events to intentional forces when they in fact were caused by unintentional forces (in other words, they could have committed the error of false-positive), or they could have attributed the events to unintentional forces when they in fact were caused by intentional forces (they could have committed the error of false-negative).[9] The consequences of false-positive errors were that our ancestors became unnecessarily paranoid and looked for predators and enemies where there were none. The consequences of false-negative errors were that our ancestors were attacked and killed by the predator or the enemy when they least expected an attack. The consequences of committing false-negative errors are much more seriously detrimental to survival and reproductive success than the consequences of committing false-positive errors, and thus evolution should favor psychological mechanisms that predispose their carriers to over-infer intentions and agency behind potentially harmless phenomena caused by inanimate objects. Evolutionarily speaking, it's good to be paranoid, because it might save your life.[10]

Different theorists call this innate human tendency to commit false-positive errors rather than false-negative errors (and as a consequence be a bit paranoid) "animistic bias"[11] or "the agency-detector mechanism."[12] These theorists argue that the evolutionary origins of religious beliefs in supernatural forces come from such an innate bias to commit false-positive errors rather than false-negative errors. The human brain, according to them, is biased to perceive intentional forces behind a wide range of natural physical phenomena, because the costs of committing false-negative errors are much greater than the costs of committing false-positive errors. It predisposes us to see the hand of God at work behind natural, physical phenomena whose exact causes are unknown.[13]

Some readers may recognize this argument as a variant of "Pascal's wager." The seventeenth-century French philosopher Blaise Pascal (1623–1662) argued that given that one cannot know for sure if God exists, it is nonetheless rational to believe in God. If one does not believe in God when He indeed exists (false-negative error), one must spend eternity in hell and damnation, whereas if one believes in God when he actually does not exist (false-positive error), one only wastes a minimal amount of time and effort spent on religious services. The cost of committing the false-negative error is much greater than the cost of committing the false-positive error. Hence, one should rationally believe in God.

### *In Church and on the Dance Floor*

More interestingly, if you have read "He Said, She Said: Why Do Men and Women Perceive the Same Situation Differently?" in chapter 3, you may see a clear parallel between the evolutionary psychological explanations of the origins of religious beliefs and Haselton and Buss's error management theory,[14] as does Haselton herself.[15] The intriguing suggestion here is that we may believe in God and the supernatural for the same reasons that men over-infer women's sexual interest in them while women under-infer men's sexual interest in them. Both religious beliefs and sexual miscommunication are consequences of the human brain designed for efficient error management, to minimize the total costs (rather than the total numbers) of committing false-positive and false-negative errors. We may believe in God for the same reason that women have to keep slapping men to set them straight or that sexual harassment is so rampant.

## **Q.** Why Are Women More Religious Than Men?

Apart from the practice of religion itself, there is something else about religion that is culturally universal. Women in virtually every society are more religious than men.

A worldwide survey asked more than one hundred thousand people from seventy different countries and regions the following two questions: "Do you believe in God" and "Independent of whether you go to church or not, would you say you are a religious person, not a religious person, or a convinced atheist?" By these measures, with only a couple of minor exceptions,[16] women in all nations and regions are more religious than men.

The sex differences in religiosity are greater in some countries (Russia) than in others (US). It is present in societies with very high levels of religiosity (Ghana, Poland, Nigeria) and in those with very low levels of religiosity (China, Japan, Estonia). It is present in all six populated continents, regardless of the particular religion involved (Catholicism in Italy and Spain, Protestantism in Germany and Sweden, Russian Orthodox in Russia and Belarus, Islam in Turkey and Azerbaijan, Shintoism in Japan, indigenous religions in Ghana, and even official atheism in China). Women are more religious than men in virtually every society surveyed. Nor is this a contemporary phenomenon. Historical records show that the sex differences in religiosity existed throughout history.[17]

Why is this? Why are women more religious than men in virtually all cultures and throughout history? What explains the universal sex difference in religiosity?

As with all other sex differences, the Standard Social Science Model offers a blanket explanation of "gender socialization." Social scientists in the Standard Social Science Model tradition contend that women are socialized to be nurturing and submissive, qualities that make religious acceptance and commitment more likely.[18] Similarly, they argue that the role of the mother subsumes religiousness, since it involves such activities as teaching the children morality and caring for the physical and spiritual welfare of other family members.[19] Some even argue that women are more religious than men because they do not traditionally work outside the home and therefore have more free time to pursue and practice religion.[20]

Unfortunately for the Standard Social Science Model, however, it turns out that there is not much empirical support for these explanations for the sex difference in religiosity. Women are more religious than men both in traditional societies, where women receive strict gender socialization, and in modern societies, where women are not subject to such strict gender socialization;[21] the experience of child rearing appears unrelated to a woman's religiosity;[22] career women are just as religious as housewives, and both are far more religious than men.[23] The preponderance of empirical evidence is therefore contrary to the Standard Social Science Model explanation for the sex difference in religiosity in terms of gender socialization.

### *Another Case of Risk Management*

The sex difference in religiosity directly follows from the evolutionary psychological theory of the origins of religious beliefs (see "Where Does Religion Come From?" above) and the sex difference in risk taking (see "Why Are Almost All Violent Criminals Men?" in chapter 6). You'll recall that the evolutionary origins of religiosity are in risk management; it is less risky to over-infer agency and hence be susceptible to religious beliefs. It is an error-management strategy to minimize the total costs of errors by predisposing the human brain to commit more false-positive errors than false-negative errors when the former has less costly consequences than the latter.[24] You'll recall, too, that women are inherently more risk-averse than men, both because women benefit far less from taking risks (given that there is a limit to how many children women can have and that all women are more or less guaranteed to have some children in their lifetime)[25] and because their offspring suffer if women are risk-seeking.[26] If men are more risk-seeking than women, and if religion is an evolutionary means to minimize risk, then it naturally follows that women are more religious than men.

Consistent with this explanation, studies show that an individual's risk preference is strongly related to his or her religiosity both across and within the sexes. Not only are women more risk-averse and more religious than men, but more risk-averse men are more religious than more risk-seeking men, and more risk-averse women are more religious than more risk-seeking women.[27] Further, consistent with this explanation, the sex difference in religiosity is larger in societies where being nonreligious is considered risky (such as in fundamentalist Christian or Muslim societies) than in societies with greater religious freedom, where individuals can freely choose to be religious or not. The sex difference is also smaller in societies where there is no widespread belief that nonbelievers go to hell, such as Buddhist societies.[28]

In the previous section of this chapter (see "Where Does Religion Come From?"), we present the intriguing possibility that humans may believe in God and the supernatural for the same reasons of error management that men over-infer women's sexual interest in them and women underinfer men's sexual interest in them. Now, in our discussion in this section of the universal sex difference in religiosity, our suggestion is that women are uniformly more religious than men for the same reasons of risk preference that men are more criminal and violent in every society. Sex differences in risk preference, religiosity, and criminality are all direct consequences of sex differences in reproductive strategy. In all areas of life, it pays for men to take risks because avoiding risks has the disastrous consequence of ending up a total reproductive loser. Religion is just another area where men are more risk-seeking than women.

## Q. Why Are Most Suicide Bombers Muslim?

According to Oxford University sociologist Diego Gambetta, editor of *Making Sense of Suicide Missions*–a comprehensive history of this topical yet puzzling phenomenon–while suicide missions are not always religiously motivated, when religion is involved, it is *always* Islam.[29] Why is this? Why is Islam the only religion that motivates its followers to commit suicide missions?

The surprising answer from the evolutionary psychological perspective is that Muslim suicide bombing may have nothing to do with Islam or the Koran (except for two lines of its text). It may have nothing to do with religion, politics, the culture, the race, the ethnicity, the language, or the region. As with everything else from this perspective, it may have a lot to do with sex–or, in this case, the *absence* of sex.

What distinguishes Islam from other major world religions (Christianity and Judaism) is that it tolerates polygyny. As we explain in chapter 2 ("Why Are Men and Women So Different?"), by allowing some men to monopolize all women and altogether excluding many men from reproductive opportunities, polygyny creates shortages of available women. If 50 percent of men have two wives each, then the other 50 percent don't get any wives at all. If 25 percent of men have four wives each, then three-quarters of men don't get any reproductive opportunities and face the distinct possibility of ending their lives as total reproductive losers.

So polygyny increases competitive pressure on men, especially young men of low status, who are most likely to be left without reproductive opportunities when older men of high status marry polygynously. It therefore increases the likelihood that young men resort to violent means to gain access to mates because they have little to lose and much to gain by doing so compared to men who already have wives. Across all societies, polygyny increases violent crimes, such as murder and rape, even after controlling for such obvious factors like economic development, economic inequality, population density, the level of democracy, and world regions.[30] So the first unique feature of Islam, which partially contributes to the prevalence of suicide bombings among its followers, is polygyny, which makes young men violent everywhere. This is the first line in the Koran that partially explains it.

## *Polygyny Is Not Enough*

However, polygyny by itself, while it increases violence, is not sufficient to cause suicide bombings. Societies in sub-Saharan Africa and the Caribbean are much more polygynous than the Muslim nations in the Middle East and Northern Africa; eighteen of the twenty most polygynous nations in the world are in sub-Saharan Africa and the Caribbean.[31] Accordingly, nations in these regions have very high levels of violence, and sub-Saharan Africa suffers from a long history of continuous civil wars, *but not suicide bombings.* So polygyny itself is not a sufficient cause of suicide bombings.

The other key ingredient is the promise of seventy-two virgins waiting in heaven for any martyr in Islam. This creates a strong motive for any young Muslim men who are excluded from reproductive opportunities on earth to get to heaven as martyrs. The prospect of exclusive access to seventy-two virgins in heaven may not be so appealing to anyone who has even one mate on earth, which strict monogamy guarantees. However, the prospect is quite appealing to anyone who faces a bleak reality on earth of being complete reproductive losers because of polygyny.

From the evolutionary psychological perspective, it is the combination of polygyny (and the resulting lack of reproductive opportunities on earth) *and* the promise of a large harem of virgins in heaven that motivates many young Muslim men to commit suicide bombings. Consistent with this explanation, all studies of suicide bombers indicate that they are significantly younger than not only the Muslim population in general but also other (nonsuicidal) members of their own extreme political organizations, like Hamas and Hezbollah. And nearly all suicide bombers are single.[32]

### *Some Puzzles in the "War on Terror"*

Some of the puzzles of the current situation in Iraq and the Middle East may begin to make sense when you shed evolutionary psychological light on them. For example, the Iraqi insurgents have killed more than six times as many Iraqis as Americans (6,004 Iraqi police and military personnel plus 10,131 civilians vs. 2,466 American troops, as of January 29, 2007).[33] From the evolutionary psychological perspective, the Iraqi insurgents may be unconsciously trying to eliminate as many of their male sexual rivals (fellow Iraqi men) as possible, rather than killing American troops (the infidels and occupiers). According to Yale University political scientist Stathis N. Kalyvas, this is precisely what happened in civil wars in two other Muslim nations (Algeria and Oman).[34] While it is difficult to remember in light of the daily news reports from the occupied Iraq, insurgency has not always been a necessary response to foreign occupation throughout history. There was absolutely no insurgency against the Allied occupation after World War II either in Germany or Japan.

While Muslim suicide bombers are collectively known as "the terrorists," they are very different from traditional terrorist groups, such as the Irish Republican Army, ETA (Basque Fatherland and Liberty), the Japan Red Army, and other Marxist revolutionaries. Terrorists, traditionally, have clear political goals and are willing to resort to violence and destruction in order to achieve them. For traditional terrorists, what is most important are political goals, and violence and destruction are means to their goals. For example, while the IRA has assassinated many targeted individuals (mostly politicians and British soldiers), they do not aim to kill random civilians. That is why when the IRA sets explosive bombs on commercial targets in Britain, it usually gives a 45-minute advance warning, enough time for the occupants of the buildings to evacuate them safely, but not enough time to call the bomb squad to locate and defuse the bombs.[35] While the members of Greenpeace and other "eco-terrorist" groups often endanger their own lives, they are not known to intentionally endanger the lives of others. Traditional terrorist groups let the whole world know that they are responsible for the violence and destruction and often court media attention, because such publicity helps spotlight their political agenda.

Our enemies in the current "War on Terror" are very different. They aim to endanger as many lives as possible, including their own, and they do not seem to have clearly stated political goals.[36] They do not give advance warnings of their attacks, and they do not even publicly claim responsibility for the violence after the fact.[37] (Many of the claims of responsibility on various websites are usually false.) It appears that murder and destruction *is* the goal, rather than the means to political goals. This may be why, for example, the Palestinians did not stop their suicide bombings even when the Israeli government under Ehud Barak conceded virtually everything that the

Palestinians demanded (the total withdrawal of Israelis from the West Bank and the control of Jerusalem).[38]

Many of these puzzles begin to make more sense when you look at the situation from the evolutionary psychological perspective. Maybe these devastating suicide bombings are not "terrorist" acts, as the term is usually used. Maybe they have nothing to do with Israel or the American and British troops. Maybe they're all about sex, as everything else in life is.

## Q. Why Is Ethnic and Nationalist Conflict So Persistent throughout the World?

If you pay attention to the world news, you know that ethnic and nationalist conflict has unfortunately been a constant feature of human history. It is no exaggeration to say that there has not been a region or a historical period that has not been affected by some sort of ethnic and nationalist conflict, and this is unfortunately still true at the dawn of the third millennium of recorded human history. The history of human civilization has in large part been a history of ethnic and nationalist conflict.

Why is this? Why is ethnic and nationalist conflict so persistent throughout history and the world?

Nationalism and other forms of ethnic movement pose a puzzle—especially for a school of the Standard Social Science Model called the rational choice theory.[39] All benefits of successful nationalist or ethnic movements, such as ethnic independence, political autonomy, and state recognition, are shared equally by everyone. So, for example, once ethnic independence is granted to a nation (say, Quebec in Canada), all members of the nation are equally independent, and no one can be excluded from enjoying the newly acquired ethnic independence. It means that those members of an ethnic group or a nation who did not contribute at all toward the cause (the "freeriders") get to enjoy the benefits of successful ethnic movements as much as those who risked life and limb in order to achieve the success (the "zealots").[40] Freeriders and zealots enjoy the same level of freedom and independence. Why, then, would anybody risk injury and death in order to bring about the change? In any situation like this, it is always rational to freeride, and no rational actors will ever contribute.[41] If everyone is rational, then no one will contribute to the cause, and it will not get off the ground, let alone succeed. How, then, can any ethnic and nationalist movement ever succeed?

### *Evolutionary Psychology Is Rational Choice for the Genes*

Once again, evolutionary psychology can solve puzzles left unresolved by the Standard Social Science Model in general or the rational choice theory in particular.[42] Joseph M. Whitmeyer was a student of Pierre L. van den Berghe, whom we have encountered a couple of times earlier. Whitmeyer argues, and mathematically proves, that any gene that inclines its carriers to help others whom they might marry, or those whose children their children might marry, or those whose grandchildren their grandchildren might marry, etc., will be favored by evolution and thus spread.[43] By contributing toward the welfare of other members of such an "extended family" or tribe, so to speak, you are essentially providing benefits for your genetic offspring, both distant and near. Whitmeyer argues that what usually passes as an ethnic group is essentially such an extended family because members of ethnic groups tend to intermarry.

Whitmeyer's insight is that while it is *economically* irrational to contribute toward ethnic and nationalist movements, as the rational choice theorists point out, because the benefits of successful ethnic collective action cannot be excluded from freeriders, it is nonetheless *evolutionarily* and *biologically* rational. It is irrational from the individual's perspective; it is rational from the genes' perspective.[44]

Whitmeyer's solution to the problem of ethnic and nationalist conflict contains both good and bad news. The bad news is that our tendency toward ethnocentrism–our desire to help and promote others of "our own kind"–is probably innate. Because they assume that humans are born blank slates, social scientists have always argued that individuals are born entirely free of prejudice, but learn to be racist and ethnocentric through childhood socialization, usually by racist parents. Evolutionary psychology in general and Whitmeyer's work in particular suggest that this is unlikely to be the case.[45] Humans are instead born racist and ethnocentric, and learn through socialization and education not to act on such innate tendencies. Humans are innately ethnocentric because ethnocentrism–helping others of one's group members at the cost of all others–was adaptive in the ancestral environment.

The good news is that we can easily overcome our innate ethnocentric tendencies. A recent experiment with an incredibly ingenious design–conducted by Robert O. Kurzban, John Tooby, and Leda Cosmides–demonstrates that while we are born with fixed categories for sex and age, we are not born with fixed categories of what constitutes a race or ethnic group, or what defines "us" versus "them."[46] We will never be able to eliminate our innate ethnocentric tendencies, but we can lessen hostility and conflict between any particular set of ethnic, religious, national, or cultural groups. How? Whitmeyer's mathematical model provides the answer: intermarriage. Our brain is designed to perceive anybody within an "extended family" of intermarrying individuals as "us," and anybody outside of it as "them." If members of hostile groups began intermarrying, we could eventually eliminate the hostility itself.

Of course, this is far easier said than done. It would be very difficult to convince members from different ethnic and national groups in conflict to marry each other. But at least there is hope. Kurzban, Tooby, and Cosmides' experiment shows that humans will never be able to stop treating men and women differently, or the young and the old differently, but they will be able to stop treating Catholics and Protestants differently in Northern Ireland; Muslims and Jews differently in Israel; or Serbs, Croats, and Muslims differently in Bosnia. While some may think there is a chicken-and-egg problem here–does ethnic conflict lead to a lack of intermarriage, or does a lack of intermarriage lead to ethnic conflict?–Whitmeyer's mathematical model points to the latter answer and suggests that it is possible to reduce ethnic and nationalist conflict via increased intermarriage.

## Q. Why Are Single Women More Likely to Travel Abroad—and Why Are Young Single Men More Likely to Be Xenophobic?

Ask a group of friends what their hobbies are. If you have many young, unmarried friends of both sexes, chances are that many of your female friends would mention traveling as one of their hobbies, while very few of your young unmarried male friends would. Alternatively, you may find that many of your young single female friends have recently been to a foreign country on a vacation, but few of your young single male friends have. Why is this?

Make a completely different observation. Pay close attention to the news coverage of the most recent Ku Klux Klan rally in the

United States or the convention of the British National Party or any other gathering of an expressly xenophobic organization. You will notice that most participants in such xenophobic organizations are young, unmarried men; there are comparatively few women or older men in the membership of such organizations. Why? It turns out that the reason why more young single women vacation abroad may be the same as why most neo-Nazis are young single men. It may have to do with a zoological phenomenon called *lekking.*

*Lek* is a Swedish word for "play" and refers in zoology to a complex of behavior whereby members of one sex, almost always male, strut and display their genetic quality in a contest, in front of an audience consisting of members of the other sex, almost always female. At the end of the lek, the females choose the winner and exclusively mate with him. The winner of lekking monopolizes all of the mating opportunities, and none of the others get any.

At first sight, humans appear to be an exception in nature. Among most species, males are gaudy, colorful, decorated, and ornamented, while females are drab in appearance. (Compare peacocks with peahens.) Males of lekking species display their physical features in order to attract mates, and females choose their mates on the basis of the males' physical appearance; the gaudier and more colorful, the better. In contrast, among humans, it is women for whom physical appearance is more important for their mate value, and it is men who choose their mates mostly for their physical appearance. (See "Why Do Men Like Blonde Bombshells [and Why Do Women Want to Look Like Them]?" in chapter 3.) And, at least in industrial societies, women tend to be more decorated and ornamented than men, although men in many preindustrial societies often wear more elaborate ornamentation than do women.

The female of most species in nature does not receive any material benefit from her mates; the male does not make any parental investment beyond the sperm deposited inside the female body during copulation. This is why the male's genetic quality is especially important for the female; in fact, nothing else matters. So among these species, males display their genetic quality in lekking, and the females choose their mates solely on the basis of their genetic quality. Human males are exceptional in nature in this regard; they make a large amount of material investment in their offspring, even though they don't make as much parental investment as women do (see "Why Are There So Many Deadbeat Dads but So Few Deadbeat Moms?" in chapter 5). This does not mean, however, that their genetic quality is not important to women; men's genetic quality can predict their future ability to acquire resources and attain status, hence their ability to make parental investment.[47] For humans, because of high male parental investment, what is important is not the male's genetic quality per se but earning potential. His genetic quality is important only to the extent that it predicts or correlates with his potential to earn and accumulate material resources.

This is why when men lek, they display their earning potential and accumulated wealth in addition to their genetic quality. And unlike other lekking species, like the sage grouse or the antelope, men lek mostly by nonphysical means. They drive luxury cars, wear expensive watches and designer suits,[48] carry electronic gadgets like cell phones and PDAs, and brag about their achievements in casual conversations.[49] Young men also advertise their genetic quality and earning potential by "cultural displays"—excelling in such "quantifiable, public, and costly" activities as music, art, literature, and science.[50]

In one study, for example, researchers covertly observed patrons of a bar in central Liverpool in the late 1990s, when cell phones were still relatively rare and expensive. The researchers discovered that men's tendency to place their cell phones on the table in clear view of others, unlike women's tendency to do the same, increases with the number of men in their group and its ratio of men to women.[51] The researchers' interpretation is that men do this, consciously or unconsciously, in order to compete with other men in their group for the attention of the women, and to display their wealth and status and hence their genetic quality and earning potential. So men lek via social and cultural, rather than physical, ornamentation.[52]

### *A Not-So-Universal Language*

Such social and cultural ornamentation, however, presents men with one problem that males of other species, who lek via physical ornamentation, do not face: It does not travel well. Social and cultural ornamentation is, by definition, socially and culturally specific. Men cannot brag about their achievements in conversations with women unless they speak the same language. Yanomamö women in the Amazon rain forest would not be able to tell the difference between a BMW and a Hyundai or the difference between an Armani suit and a Burger King uniform, and their status implications; a Grammy or a Nobel Prize will not impress them at all. (Has any Nobel Prize winner ever had massive head scars, indicating their experience in club fights?) Conversely, Western women are unlikely to be impressed by body scars and large penis sheaths. Signs of men's status and mate value are specific to societies and cultures, and they lose meaning outside of them.

This is in clear contrast to women's status and mate value. Standards of youth and physical attractiveness, the two most important determinants of women's status and mate value, are culturally universal[53] because they are innate[54] (see chapter 3, "Why Is Beauty *Not* in the Eye of the Beholder or Skin-Deep?"). Men in preliterate and innumerate cultures without any concept of fractions or the decimal point will be able to distinguish between women with 1.0 and 0.7 waist-to-hip ratios. Yanomamö men will see that a Victoria's Secret lingerie model is extremely *moko dude* (a Yanomamö phrase meaning "perfectly ripe").[55]

### *A Sure Sign That Someone Wanted You*

If men's status and mate value are specific to their own society and culture, then they should avoid different cultures, where a completely different set of rules, of which they are ignorant, may apply. In contrast, women should not avoid foreign cultures to the same extent that men do, because rules applicable to them are cross-culturally universal.

However, this sex difference should disappear once men marry, for a couple of reasons. First, married men who have achieved reproductive success should have less of an urgent need to attract mates by social and cultural ornamentation than do unmarried men.[56] Second, and more important, *mates are probably the only ornamentation or lekking device men can display that is cross-culturally meaningful.* There is evidence that females of species as varied as guppies,[57] Japanese medaka,[58] black grouse,[59] and Japanese quail[60] prefer to mate with males who have recently mated. Females use other females' choice of males as evidence of their genetic quality; in other words, they copy each other. And some suggest that human females might do the same.[61]

The idea is simple: If a woman meets a strange man, she has no basis on which to form an opinion of him. He can be a high-quality man, or he can be a low-quality man; she just doesn't know. However, if he has a wife, that means that at least one woman, who presumably closely inspected his quality before marrying him, found him good enough to marry. So he couldn't be *that* bad after all; at least one woman found him desirable. So being married (the presence of a wife) is one cross-culturally transportable ornamentation or lekking device that signifies men's superior mate value, and married men should not avoid foreign cultures.[62]

Dislike of foreign cultures can be measured by the likelihood of travel to foreign countries or by the expressions of xenophobic attitudes. One empirical study with a large European sample shows that, controlling for age, education, and income (factors that are expected to, and in most cases do, affect people's ability to travel), unmarried women are significantly more likely to vacation abroad than unmarried men.[63] The same study also demonstrates that, controlling for age and education, unmarried women are significantly less likely to express xenophobic attitudes than unmarried men toward individuals of other nationalities, races, and religions. The pattern is similar among Americans as well.[64] In all cases, the sex difference disappears once the respondents are married; married women are no more likely to travel to foreign countries (probably because married couples tend to vacation together) or no less likely to express xenophobic attitudes than married men.

Both the likelihood of travel abroad and expressions of xenophobia reflect men's need to attract women using social and cultural ornamentation. Men's status and mate value, unlike women's, are socially and culturally specific, and they cannot successfully attract women outside of their own society and culture. Married men, on the other hand, can use their wives as cross-culturally meaningful social ornamentation to signify their mate value. In sharp contrast, the standards and criteria by which women are judged for their mate value are socially and culturally universal, and thus women have no need to fear foreign cultures.

Conclusion

# Stump the Evolutionary Psychologists

## A FEW TOUGHER QUESTIONS

In the preceding chapters, we have used evolutionary psychology to explain a wide variety of puzzles in many areas of social life, from sex and mating to marriage and the family; from crime and violence to economics, politics, and religion. We hope we have succeeded in convincing you that evolutionary psychology is an approach that can provide at least some (partial) answers to many persistent questions about human behavior. Given the range and number of questions that we have attempted to address in this book, you may rightfully wonder if there is anything that evolutionary psychology *cannot* address.

In the appendix of his 1994 international bestseller *The Moral Animal: The New Science of Evolutionary Psychology* (the book that introduced evolutionary psychology to both of us), the science writer Robert Wright lists six questions that evolutionary psychology could then not answer. We would like to return to these six questions and see how far evolutionary psychology has come in solving these puzzles in the last thirteen years.

1. ***What about homosexuals?***

How can evolutionary psychology explain homosexuality, when it places so much emphasis on reproductive success (accomplished largely through heterosexual sex) as the ultimate motive of all behavior? This is usually among the first questions that we receive from students, neighbors, and academic colleagues alike when we present our evolutionary psychological ideas. In his 2000 book *The Mating Mind: How Sexual Choice Shaped the Evolution of Human Nature*, the evolutionary psychologist Geoffrey F. Miller admits that he cannot explain homosexuality.[1] To the best of our knowledge, no one else can. There is still no definitive and accepted explanation of homosexuality in 2007, thirteen years after Wright posed the question "What about homosexuals?"

There are some ideas, however, not from evolutionary psychology but from the related field of behavior genetics. The geneticist Dean Hamer and his team have discovered the genetic roots of male homosexuality. While they have not discovered the "gay gene" (the actual genes that increase the probability of male homosexuality), they have located a region of a chromosome that is involved–a genetic marker at Xq28.[2]

Even if a "gay gene" or genes are eventually discovered and sequenced, it still does not explain how such genes can survive if their carriers are exclusively or predominantly homosexual. The evolutionary biologists Robert L. Trivers and William R. Rice have an idea. Their theory of the evolution of male homosexuality is described in Hamer's book *The Science of Desire: The Search for the Gay*

*Gene and the Biology of Behavior.*[3] Trivers suggested the theory to Hamer in a phone conversation, but he and Rice never published their theory.[4]

Trivers and Rice believe that genes for male homosexuality may be passed on to the next generation, not by the gay men themselves but by their sisters and other female relatives. They posit that the so-called gay genes may incline their carriers, male or female, toward the same behavior: the desire to have sex with men. If the carriers are male, then they become male homosexuals (and they do not pass on their genes to the next generation). If the carriers are female, however, they will have more children than other women who do not carry such genes because they will have a larger number of male sex partners and have sex with them more frequently. The reduced reproductive success of gay men is compensated for by the heightened reproductive success of their sisters, and the male homosexuality genes will survive.

Their idea, which might be dubbed the "horny sister hypothesis," is a very provocative one, but it has been supported by a recent study that shows that female maternal relatives of male homosexuals have more children than female maternal relatives of male heterosexuals.[5] (Because they are on the X chromosomes, the "gay genes" are passed on to gay men by their mothers, not their fathers.)

We also note, however, that the most likely reason that male homosexuality has survived to this day is that throughout most of recorded history, gay men were forced to hide their sexuality by social norms and legal sanctions, and so got married and had children like straight men.[6] If so, the liberation of homosexuals, which allows them to come out of the closet and not pretend to be straight, may ironically contribute to the end of homosexuality.

So far no one has found a genetic basis for female homosexuality.

2. ***Why are siblings often so different from one another?***

This is one of the clear examples of the triumph of recent evolutionary psychological theory and research. This was an unresolved question in 1994 but has now been answered, thanks to two somewhat unconventional heroes of evolutionary psychology: Frank J. Sulloway[7] and Judith Rich Harris.[8]

In his 1996 book *Born to Rebel: Birth Order, Family Dynamics, and Creative Lives*, Frank J. Sulloway argues that siblings within the same family must occupy different familial niches.[9] Firstborns (the eldest siblings), who are born into a family without any siblings with whom to compete for parental resources, typically grow up to identify with the parents and, by extension, other authority figures. Later-borns (younger siblings), in contrast, are always born into a family in which there are older siblings who have already taken the niche of identifying with the parents. So they must carve out their own niche by distancing themselves from the parents and becoming rebels. Sulloway's massive historical data show that in religious, political, and scientific spheres, firstborns are more likely to become the conservative vanguards of the old tradition, and laterborns are more likely to become the leaders of new revolutions. Thus, birth order, whether one is the eldest or younger, is one factor that makes siblings raised in the same family different in their personalities.

In her 1998 book *The Nurture Assumption: Why Children Turn Out the Way They Do*, Harris methodically demolishes the universally held assumption that how parents raise their children is a major determining factor in how they turn out.[10] Harris instead argues that parental socialization has very little effect on children because they are mostly socialized and influenced by their peers. While Harris's conclusion was enormously controversial and widely condemned by politicians and the media alike, it is in fact corroborated by behavior genetic research.[11] Behavior geneticists contend that the rough rule of thumb when it comes to the determinants of child development is 50-0-50—that is, roughly 50 percent of the variance in personality, behavior, and other traits is heritable (determined by the genes); roughly 0 percent by shared environments (what happens within the family, shared by all siblings); and roughly 50 percent by the nonshared environment (what happens outside the family, often not commonly shared by siblings).

Harris's work highlights the importance of the nonshared environment on child development, and therefore gives us another reason why siblings raised in the same family are different. Of course, contrary to how the media portrayed (and viciously attacked) Harris's work, it decidedly does *not* mean that parents are not important for children's development. Parents are *enormously* important because children receive 100 percent of their genes from their biological parents. It simply means that within broad limits, how parents socialize their children is not very important to adult personality.

Both Sulloway's and Harris's research and conclusions have been hugely controversial, but mostly in the media and among nonacademic audiences. The scientific community in general and evolutionary psychologists in particular tend to be supportive of their ingenious research and counterintuitive conclusions. Because Sulloway's theory emphasizes family dynamics as a primary determinant of adult personality, whereas Harris's group socialization theory focuses on what happens outside the family, Sulloway and Harris are naturally critical of each other's work.[12]

3. ***Why do people choose to have few or no kids?***
4. ***Why do people commit suicide?***

In contrast to the success of evolutionary psychology and behavior genetics in solving the first two questions on Wright's 1994 list, the next two questions present us with still unresolved puzzles. As far as we know, there is no compelling evolutionary psychological explanation for why some people choose to remain childless or why people commit suicide. Behavior geneticists have recently discovered the genetic basis for fertility behavior;[13] we now know that whether one has many children or a few is partially influenced by genes. However, this still does not explain why some people choose to have none; clearly, genetic tendency toward childlessness could not be selected for.

In some ways, the genetic influence on how many children people have goes against evolutionary logic. It suggests that people who have many siblings (because their parents had many children) themselves have many children, and people who have few siblings (because their parents had few children) themselves have few children. The evolutionary logic would suggest the opposite. If you have many siblings, you don't have to have many children yourself because you can still attain great reproductive success by investing in your siblings; both your children and your siblings carry half your genes. In contrast, if you have few siblings, you have to have many children yourself because you don't have the option of investing in your siblings. The heritability of family size therefore remains a puzzle for evolutionary psychology.

5. ***Why do people kill their own children?***

We do not know why this question was on Wright's 1994 list, because Martin Daly and Margo Wilson had already solved it in their

1988 book *Homicide*, which is partly based on their even earlier work.[14] Daly and Wilson first point out that the answer to the question, "Why do people kill their own children?" is, "They don't." Most parents who are convicted of killing their children are actually stepfathers, who are not genetically related to the children they kill. Crime statistics make it appear as though some biological parents kill their own children, because the police, uninformed by evolutionary psychology, make no distinction between biological parents and stepparents in their statistics.

From the evolutionary psychological perspective, it makes perfect sense for stepparents to neglect; underinvest in; and, in some cases, even kill their stepchildren, so that their spouses will focus their investment of time and resources in the couple's common genetic children, current and future. Among many species, such as baboons and lions, when a new male takes over a group of females with young, the first thing he does is kill all the existing children systematically, so that all the females will reproduce with him. It would be surprising if humans were any different.

Even the few cases where biological parents kill their genetic children can be explained by Daly and Wilson's notion of "discriminative parental solicitude."[15] They point out that all parents have limited resources to invest in their children. Their task is to maximize their reproductive success–not by maximizing the number of children but by maximizing the number of grandchildren. From this strictly Darwinian perspective, any resources invested in children who are not likely to survive to sexual maturity or find mates and reproduce themselves are entirely wasted. Thus, parents are far more likely to neglect, abuse, and kill their biological children who are deformed, handicapped, ill, or even physically unattractive and to shift their parental investment of their limited resources toward those children with more promising reproductive prospects.[16] As uncomfortable as we may be with such a conclusion, the truth appears to be that parents do favor some of their children over others, even among their own genetic children, to say nothing about stepchildren to whom they are not genetically related, and they overwhelmingly favor those who are intelligent, beautiful, healthy, and sociable.

6. ***Why do soldiers die for their countries?***

To the best of our knowledge, there is no satisfactory explanation for this phenomenon from an evolutionary psychological perspective. However, we have a few observations. First, it seems to us that soldiers die for their countries only if their country honors fallen soldiers and provides for their widows and children; in fact, we cannot think of any civilized society that does not honor men who fight and die for their country. Second, many men get married right before they are shipped to war, which explains why so many deployed soldiers have newborn babies whom they have not seen. Perhaps fighting and potentially dying for their countries when the country honors the war dead and provides for their widows and children is some men's reproductive strategy to make sure that their children are taken care of, when they could not provide sufficient parental investment themselves. Historically, soldiers have disproportionately come from lower classes. We may also add that while the last thing men do before they go to war is to get married and impregnate their new brides, unfortunately, sometimes the first thing they do when they conquer their enemy is to rape the women in the conquered society,[17] which also helps increase their reproductive success further. Soldiers would not get this opportunity unless they were willing to fight and die for their countries. But we admit that these are simply our observations, and we do not have a clear explanation for why soldiers die for their countries. Neither does anyone else.

So the scorecard for evolutionary psychology, as of 2007, is 3–3. Three of the six questions on Wright's 1994 list have been satisfactorily solved; the other three remain unsolved.

To this list of three remaining questions, we would like to add a few more. Here are a few other questions that currently present theoretical puzzles to evolutionary psychology.

7. ***Why do children love their parents?***

At first glance, this question may appear absurd. *Of course* children love their parents; it is only natural. But why?

If you really think about it, there is absolutely no evolutionary psychological reason why children should love and care for their parents. Some people (usually parents themselves) have suggested to us that parents will be more motivated to invest in children who love them back. But this is not true; from an evolutionary psychological perspective, parents *have to* love their children, whether the children love them back or not, in order to motivate their parental investment. And, as Daly and Wilson's work on discriminative parental solicitude shows, parents are motivated to invest not necessarily in the children who love them most, but in those who have the greatest potential to attain higher reproductive success themselves (more attractive, more intelligent, healthier children, or boys if the parents are wealthy, girls if they are poor, etc.). If parents with limited resources have two children, one an intelligent, physically attractive, and healthy child who hates them, and the other a handicapped, unattractive, and sickly child who loves them, the cold evolutionary logic dictates that the parents invest in the former, not the latter. So the children do not really have to love their parents.

This is especially true for adult children. While the parents are still young, they can potentially produce further offspring with whom the children share half their genes. So it might make sense for the children to invest in and take care of their parents, so that they can produce more siblings. But once the parents are past the reproductive age, this is no longer possible. So it makes no evolutionary psychological sense for adult children to take care of their elderly parents.

Yet the overwhelming evidence from most human societies shows that children do love their parents, and this is a theoretical puzzle for evolutionary psychology–although probably *only* for evolutionary psychology.

8. ***Why do parents in advanced industrialized nations have so few children?***

This is slightly different from question 3 above on Wright's list about why some people choose to be childless. Most people–for example, 90 percent of contemporary Americans–have children. However, despite the fact that most middle-class Americans could comfortably raise four or five children and invest sufficient resources in each of them, most parents choose to have only about two children.

In fact, there is an additional layer to this puzzle. Most Americans prefer to have a boy and a girl, rather than two boys or two girls. Parents who have two children of the same sex are more likely to have a third child than parents who have a boy and a girl.[18] Why Western parents do not have as many children as they can safely afford and invest in, and why they have a preference for a child of each sex, remains a mystery from the evolutionary psychological perspective.

9. ***Why do people find a tan attractive? Why do men hog the remote control and typically channel surf much more than women? Why are men mostly responsible for barbecuing and carving meats while women do most of the other cooking?***

These are some of the trivial observations that we and others have made that are too widespread, consistent, and strong to be coincidental or the result of cultural socialization. It is likely that there are some biological or evolutionary reasons behind such consistent observations. For example, some argue that tanning is young, single women's means to advertise their health and beauty and that this is why young, single women are more likely to get a tan than others.[19] To the best of our knowledge, no one has begun to propose explanations for any of the other puzzles.

In order for evolutionary psychology to explain such phenomena, however, we must first make sure that they are truly culturally invariant human universals (or if there are minor cultural variations, they can be explained as interactions between evolved psychological mechanisms and local ecology and environment).[20] If the observations are not truly culturally universal, then it is unlikely that they have biological or evolutionary roots. The first thing evolutionary psychologists must do in order to explain widespread behavior is to establish that it is culturally universal.

This is exactly what the pioneer evolutionary psychologist David M. Buss did in the 1980s, when he conducted research in thirty-seven different cultures on all continents to ascertain that the preferences for ideal mates expressed by students at the University of Michigan are indeed widely shared by people in all human societies.[21] Others have followed in Buss's footsteps. One recent study was conducted in fifty-two nations, ten major world regions, six continents, and thirteen islands, and found that expressions of sexual desires (such as the desire for sexual variety) and their consequences (such as the practice of "mate poaching"–stealing someone else's mate) are more or less the same in all societies.[22] Cultural universality is one of the hallmarks of the evolved mind.

We hope this quick survey of remaining theoretical puzzles in evolutionary psychology makes it abundantly clear that there are still many questions to be asked and many puzzles to be solved in modern evolutionary psychology. We both left sociology and became evolutionary psychologists in response to *one sentence* by Robert Wright in his 1994 book *The Moral Animal*: "For now, this is the state of evolutionary psychology: so much fertile terrain, so few farmers."[23] We became farmers and started tilling the fertile terrain. Wright's observation for evolutionary psychology is much less true now in 2007 than it was in 1994, as a large number of young and talented graduate students in psychology, anthropology, and elsewhere choose to pursue evolutionary psychology, which is probably the fastest growing academic field today.

At the 17th Annual Conference of the Human Behavior and Evolution Society (the main academic organization of evolutionary psychologists) held in Berlin in 2004, the then HBES president Bobbi S. Low remarked that the number of people who were on the program committee, which successfully planned and organized the Berlin conference in 2004, was larger than the entire group of people who originally gathered only two decades earlier to form the academic organization that later became HBES. Many attendees of these early meetings slept on the floor of Low's house when they met at the University of Michigan in the 1980s.[24] Twenty years later, the 2005 HBES meetings were held in the Hyatt Regency in Austin, Texas, with nearly five hundred participants. We have a feeling that five hundred house guests would have stretched even Low's enormous hospitality.

The growth of evolutionary psychology has also been international; in addition to the United States and the United Kingdom, where scientific research in every field is most active, evolutionary psychology has a particularly and disproportionately large following in Japan and Belgium. Yet we could always use more bright minds to help us solve the remaining theoretical puzzles in evolutionary psychology. Apply within.

# Afterword

In the hardcover edition of *Why Beautiful People Have More Daughters*, we ask and answer twenty-eight different questions in all areas of life. Of course, there are many more questions that evolutionary psychology can shed light on, as well as many others that still defy it. For example, there is a phenomenon that has mystified scientists for more than half a century. Everybody knew about it, but nobody knew why. Ever since I learned about it in 2000, I always wanted to solve the puzzle myself. It's also a timely question, given that we are in the middle of a war right now. The puzzle is . . .

## Q. Why Are More Boys Born during and after Major Wars?

The phenomenon was first noticed in 1954 with regard to white children born during World War II in the United States.[i] It has since been replicated for most of the belligerent nations in both world wars.[ii] The phenomenon has been dubbed the "returning soldier effect."[iii] There is no doubt that the phenomenon is real, but nobody has been able to explain it. Why are soldiers who return from wars more likely to father sons than other men?

There is now evidence that, at least among the British soldiers who fought in World War I, those who survive battle and return home to be reunited with their wives are taller than those who die, never to have another chance to have a child.[iv] A comparison of the physical characteristics of the British soldiers who survived or died in World War I shows that surviving soldiers are on average nearly one inch taller than fallen soldiers. The average height of the surviving soldiers is 66.4 inches, while that of the fallen soldiers is 65.5 inches. Even in the small sample that is examined, this one-inch difference is highly statistically significant.

As we note on page 102 ("Boy or Girl? What Influences the Sex of Your Child?" in chapter 5), taller parents are more likely to have sons than shorter parents are. So the excess boys born during and immediately after the world wars might be a consequence of the fact that taller soldiers, who are more likely to have sons to begin with, are more likely to survive the war and return home, whereas shorter soldiers, who are more likely to have daughters, are less likely to survive the war and return home to have daughters.

Now you may be asking yourself, What difference can such a small height advantage–slightly less than one inch–possibly make? Detailed calculations show that the one-inch difference is more than twice as sufficient to account for all the excess boys born in the United Kingdom during and immediately after World War I. True, a one-inch increase in height only increases the odds of having a son by 5 percent. However, because so many men (nearly one-third of those between the ages of 15 and 40 in the United Kingdom) were mobilized during World War I, the 5 percent increase in the odds of having a son for the taller surviving soldiers translates into millions of excess boys. It is more than enough to account for the entire "returning soldier effect" in the United Kingdom.

### *But Why Are Taller Soldiers More Likely to Survive Battle?*

This particular explanation of the "returning soldier effect" leads to another question: Why do taller soldiers have a greater chance of survival in war? This is still a puzzle, and I don't have a definitive answer. But there are some possibilities.

First, taller soldiers, especially during the less prosperous times of the early twentieth century, may have been physically stronger and more fit, as well as possibly genetically and developmentally healthier. So they might have been better able to resist diseases and wounds sustained during combat, which might have killed their shorter and less healthy comrades.

Second, height is known to be correlated with intelligence.[v] Although scientific opinions vary as to *why* taller people are more intelligent than shorter people, the fact that they are is beyond dispute. If taller soldiers on average are more intelligent than shorter soldiers, then they may be expected to achieve higher ranks within the military. Even though the sample used in the study of the British soldiers in World War I includes only enlisted men and noncommissioned officers and excludes commissioned officers, it is possible that taller and thus more intelligent soldiers were able to climb the ranks of noncommissioned officers to such ranks as lance corporal and sergeant, and were able to avoid the most dangerous combat situations because of their relative rank. Alternatively, taller and more intelligent soldiers might have been better able to fight successfully and survive in modern wars. For example, a surprising number of British soldiers survived World War I by deserting. They may have needed higher intelligence to desert and avoid court-martial successfully.

Finally, my colleague Dominic D. P. Johnson at the University of Edinburgh has made a very interesting suggestion to me. Vital organs in the body may not grow in size in exact proportion to the body. In other words, taller soldiers may have bigger vital organs such as the heart and lungs, but they may not be as big as they should be given their body size. If this is the case, then bigger soldiers, while they are statistically more likely to be shot because of their larger body size, nonetheless have more room in their body where they can be "safely" shot and still survive the injury.

### *Are There More Boys Being Born during the Current War?*

Now does this mean that more boys are and will be born during and after our current war? Not likely.

Regardless of the exact reason that taller soldiers are more likely to survive battle, the phenomenon of the "returning soldier effect" is not likely to be observed and repeated in more recent and future wars. This is because a substantial proportion of the population must be deployed for the proposed mechanism to produce excess boys in the population. Military forces of advanced Western nations today do not require as many soldiers as they used to. The transition to smaller military forces is reflected in the discontinuation of a mandatory draft in most Western nations.

With much smaller proportions of the population mobilized in wars, the returning soldier effect is not likely to be repeated, even if taller soldiers are still more likely to survive battle and even if taller parents are more likely to have sons. The higher offspring sex ratios among surviving (and returning) soldiers will not significantly shift the offspring sex ratio of the whole society. Even though an increasing number of young men and women are mobilized in the current war, the rate of mobilization in the United States is nowhere near one-third. Probably for this reason, more boys were *not* born during more recent wars, such as the Iran-Iraq wars in 1980–1988[vi] and the ten-day war in Slovenia in 1991.[vii] Nevertheless, if the height advantage of surviving soldiers over fallen soldiers in the United Kingdom during World War I is generalizeable to other belligerent nations in both world wars, then this can potentially solve one of the long-standing mysteries in evolutionary psychology.

–Satoshi Kanazawa
London, February 2008

## Sources


i. MacMahon, Brian and Thomas F. Pugh. 1954. "Sex Ratio of White Births in the United States during the Second World War." *American Journal of Human Genetics.* 6: 284–92.

ii. Graffelman, Jan and Rolf F. Hoekstra. 2000. "A Statistical Analysis of the Effect of Warfare on the Human Secondary Sex Ratio." *Human Biology.* 72: 433–45.

iii. Cartwright, John. 2000. *Evolution and Human Behavior: Darwinian Perspectives on Human Nature.* Cambridge: MIT Press.

iv. Kanazawa, Satoshi. 2007. "Big and Tall Soldiers Are More Likely to Survive Battle: A Possible Explanation for the 'Returning Soldier Effect' on the Secondary Sex Ratio." *Human Reproduction.* 22: 3002–8.

v. Case, Anne and Christina Paxson. 2006. "Stature and Status: Height, Ability, and Labor Market Outcomes." NBER Working Paper Series 12466. Available online at www.nber.org/papers/w12466. Accessed February 29, 2008.
Jensen, Arthur R. and S. N. Sinha. 1993. "Physical Correlates of Human Intelligence." Pp. 139–242 in *Biological Approaches to the Study of Human Intelligence*, edited by Philip A. Vernon. Norwood, NJ: Ablex.

vi. Ansari-Lari, M. and M. Saadat. 2002. "Changing Sex Ratio in Iran 1976–2000." *Journal of Epidemiology and Community Health.* 56: 622–23.

vii. Zorn, Branko, Veselin Sucur, Janez Stare, and Helena Meden-Vrtovec. 2002. "Decline in Sex Ratio at Birth After 10-Day War in Slovenia." *Human Reproduction.* 17: 3173–7.

# Notes

### Preface

1 Miller and Kanazawa (2000).

### Introduction

1 Maynard Smith (1997).
2 Ridley (1999, pp. 54–64).
3 Pinker (2002).
4 Scarr (1995).
[5] Wilson (2007) is a late exception.
6 Moore (1903).
7 Hume (1739).
8 Davis (1978).
9 Ridley (1996, pp. 256–8).
10 Alexander et al. (1979); Kanazawa and Novak (2005).
11 Calden, Lundy, and Schlafer (1959); Gillis and Avis (1980); Sheppard and Strathman (1989).
12 Davis et al. (1993); Rand and Kuldau (1990).

### Chapter 1

1 Buss (1989); Daly and Wilson (1988).
2 Barkow, Cosmides, and Tooby (1992).
3 Ellis and Bjorklund (2005, p. x).
4 Barkow (2006); Tooby and Cosmides (1992, pp. 24–49).
5 Ellis (1996); Daly and Wilson (1988, pp. 152–6).
6 Campbell (1999, p. 243).
7 Pinker (2002).
8 Ellis (1996).
9 Cornwell, Palmer, and Davis (2000); Cornwell et al. (2001); Machalek and Martin (2004).

[10] In this section we present a brief introduction to the field of evolutionary psychology. Nonacademic readers who would like a more extensive introduction may consult David M. Buss's *The Evolution of Desire: Strategies of Human Mating* (Buss 1994), Matt Ridley's *The Red Queen: Sex and the Evolution of Human Nature* (Ridley 1993), and Robert Wright's *The Moral Animal: The New Science of Evolutionary Psychology* (Wright 1994). Academic readers might want to consult Barkow, Cosmides, and Tooby (1992); Buss (1995, 1999); Cartwright (2000); Daly and Wilson (1988); and Kanazawa (2001a).

[11] We must define natural and sexual selection explicitly, because our usage may appear a bit unorthodox to anyone with some background in evolutionary biology. We define *natural selection* as the process whereby some individuals live longer than others, and *sexual selection* as the process whereby some individuals leave more offspring (or copies of their genes) than others. Natural selection is a matter of survival; sexual selection is a matter of reproductive success. This is how Darwin originally defined natural and sexual selection–as two separate processes. That is why he wrote two separate books–*On the Origin of Species by Means of Natural Selection* (1859) to explain natural selection, and *The Descent of Man, and Selection in Relation to Sex* (1871) to explain sexual selection. In the 1930s, however, biologists redefined natural selection to subsume sexual selection and began to contend that differential reproductive success was the currency of natural selection. This is now the orthodoxy in all biological textbooks, which claims that sexual selection is but one branch of natural selection (Cronin 1991, pp. 231–43).

In this book, we argue against this orthodoxy. We concur with Geoffrey F. Miller (2000, pp. 8–12), Anne Campbell (2002, pp. 34–5), and others in the current generation of evolutionary psychologists and believe that we should return to Darwin's original definitions and treat natural and sexual selection as two distinct processes. This is still controversial and of the minority, but we firmly believe that the conceptual separation of natural and sexual selection will bring clarity in evolutionary biology and psychology.

12 Williams (1966).

13 Barash (1982, pp. 144–7).

14 Daly, Wilson, and Weghorst (1982).

15 Gaulin, McBurney, and Brakeman-Wartell (1997).

16 Cerda-Flores et al. (1999).

17 Baker and Bellis (1995, p. 200, box 8.4).

18 Buss (1988, 2000); Buss and Shackelford (1997).

19 White (1981); Buunk and Hupka (1987).

20 Buss, Larsen, and Westen (1992); Buss et al. (1999).
21 Harris (2003); DeSteno et al. (2002).
[22] Pietrzak et al. (2002). As one of the deans of modern evolutionary psychology, David M. Buss, points out (Buss, Larsen, and Western 1996), evolutionary psychologists (Daly, Wilson, and Weghorst 1982; Symons 1979, pp. 226–46) *predicted* the existence of these sex differences in romantic jealousy on the basis of evolutionary logic alone more than a decade before any systematic data existed.
23 Betzig (1997a).
24 van den Berghe (1990, p. 428).
25 Endorsement on the cover of Betzig (1997b).
26 Bowlby (1969).
27 Kanazawa (2002, 2004b).
28 Kanazawa (2002).
29 Buss (1988).
30 Crawford (1993); Symons (1990); Tooby and Cosmides (1990).
31 Kanazawa (2004a).

## Chapter 2

1 Blum (1997); Mealy (2000); Moir and Jessel (1989); Pinker (2002, pp. 337–71).
2 Connellan et al. (2000).
3 Alexander and Hines (2002).
4 Brown (1991); Pinker (2002, appendix, pp. 435–9).
5 Alexander et al. (1979); Daly and Wilson (1988, pp. 140–2).
[6] Actually, as we argue in chapter 4, all human societies at all times are polygynous. They practice either simultaneous polygyny, allowing some men to have multiple wives simultaneously, as happens in Muslim and African tribal societies; or serial polygyny, allowing some men to have multiple wives sequentially through divorce and remarriage. The only truly and strictly monogamous societies prohibit simultaneous polygyny, divorce and remarriage, or extramarital affairs. No human societies known to anthropologists belong to this category (Betzig 1989) and thus more men than women always remain mateless in every society, given a roughly 50–50 sex ratio.
[7] The figure most often cited for the total number of children sired by Moulay Ismail the Bloodthirsty, taken from the 1976 edition of the *Guinness Book of World Records* (McWhirter and McWhirter 1975), is 888. However, according to the 1995 edition of the same book (Young 1994, p. 10), it is at least 1,042.

8 Betzig (1986).
9 Campbell (1999, 2002).
10 Clutton-Brock and Vincent (1991).
11 Trivers (1972).
12 Harris (1974).
13 Chomsky (1957).
14 Pinker (1994).
15 van den Berghe (1990, p. 428).
16 Freeman (1983, 1999).
17 Chagnon (1968).
18 Nance (1975).
19 Hemley (2003).
20 Ridley (1996, pp. 213–5).

## Chapter 3

1 Buss (1989).
2 Bloch (1994, pp. 1–13).
3 Abdollahi and Mann (2001).
4 Crawford, Salter, and Jang (1989).
5 Kanazawa and Still (2000b).
6 Etcoff (1999, pp. 89–129); Mesko and Bereczkei (2004).
7 Etcoff (1999, pp. 122–6).
8 Singh (1993); Singh and Young (1995); Singh and Luis (1995).
9 Jasienska et al. (2004).
10 Symons (1995, p. 93).
11 Cartwright (2000, pp. 153–4).
12 Marlowe (1998).
13 Jasienska et al. (2004).
14 Rich and Cash (1993).
15 Bloch (1994, pp. 1–13).
16 Wall (1961).
17 Ramanchandran (1997).
18 Feinman and Gill (1978).
19 van den Berghe and Frost (1986).
20 Wong and Ellis (1984).
21 Feinman and Gill (1978).
22 Ridley (1993, pp. 293–5).
23 Feinman and Gill (1978).

24 Kenrick and Keefe (1992).
25 Hess (1975); Hess and Polt (1960).
26 Feinman and Gill (1978, p. 47, table 1).
27 Cunningham, Druen, and Barbee (1997).
28 Wagatsuma and Kleinke (1979).
29 Bernstein, Tsai-Ding, and McClellan (1982); Cross and Cross (1971).
30 Cunningham et al. (1995).
31 Jones (1996); Jones and Hill (1993).
32 Maret and Harling (1985).
33 Morse and Gruzen (1976).
34 Thakerar and Iwawaki (1979).
35 Langlois et al. (1987); Samuels and Ewy (1985).
36 Slater et al. (1998).
37 Langlois, Roggman, and Rieser-Danner (1990).
38 Symons (1995).
39 Little et al. (2002).
40 Gangestad, Thornhill, and Yeo (1994); Mealey, Bridgstock, and Townsend (1999); Perrett et al. (1999).
41 Bailit et al. (1970); Møller (1990, 1992); Parsons (1992).
42 Parsons (1990).
43 Gangestad and Buss (1993).
44 Langlois and Roggman (1990); Rubenstein, Langlois, and Roggman (2002).
45 Langlois and Roggman (1990).
46 Thornhill and Gangestad (1993).
47 Thornhill and Møller (1997, pp. 528–33).
48 Langlois et al. (2000); Shackelford and Larsen (1999).
49 Hönekopp, Bartholomé, and Jansen (2004).
50 Henderson and Anglin (2003).
51 Al-Eisa, Egan, and Wassersub (2004).
52 Kalick et al. (1998).
53 Grammer and Thornhill (1994).
54 Langlois et al. (1994).
55 Trivers (1972).
56 Pérusse (1993, pp. 273–4).
57 Pérusse (1993, p. 273).
58 Buss and Schmitt (1993).
59 Ellis and Symons (1990).
60 Buss and Schmitt (1993).

61 Salmon and Symons (2001, 2004).
62 Symons (1979, pp. 170–84).
63 Ellis and Symons (1990).
64 Buss and Schmitt (1993).
65 Hejl, Kammer, and Uhl (Forthcoming).
66 Carroll (1999); Gottschall et al. (2004); Thiessen and Umezawa (1998); Wilson (1998, pp. 210–37).
67 Buss (1989).
68 Kenrick and Keefe (1992).
69 Kenrick and Keefe (1992).
70 Abbey (1982).
71 Kanazawa (2006a); Liedtke (2000); Pate (2001); Ream (2000).
72 Haselton (2003); Haselton and Buss (2000).
73 Yamagishi, Jin, and Kiyonari (1999).
[74] If you are familiar with elementary statistics, you recognize the false-positive and false-negative errors as "Type I" and "Type II" errors.
75 Haselton and Buss (2000, p. 90); Haselton and Nettle (2006).
76 Yamagishi et al. (Forthcoming).
77 Guthrie (1993).
78 Boyer (2001).

## Chapter 4

1 Emlen (1995).
2 Smith (1984, p. 604, figure 1A).
3 Smith (1984, p. 609).
4 Cerda-Flores et al. (1999); Gaulin, McBurney, and Brakeman-Wartell (1997).
5 White (1988).
6 Chisholm and Burbank (1991).
7 Bellis and Baker (1990); Birkhead and Møller (1991).
8 Cartwright (2000, p. 222, table 8.1).
9 Gallup et al. (2003, p. 278).
10 Gallup et al. (2003, p. 278).
11 Gallup et al. (2003).
12 Baker and Bellis (1995).
13 Barash and Lipton (2001).
14 Alexander et al. (1979); Leutenegger and Kelly (1977).
15 Alexander et al. (1979, pp. 428–30, table 15-3).

16 Mealey (2000, p. 306).
17 Alexander et al. (1979); Leutenegger and Kelly (1977).
18 Kanazawa and Novak (2005).
19 Silventoinen et al. (2001).
20 Biro et al. (2001); Frisch and Revelle (1970); Helm, Münster, and Schmidt (1995); Jaruratanasirikul, Mo-suwan, and Lebel (1997); Nettle (2002); Okasha et al. (2001).
[21] There is a third, even newer explanation of the evolution of sexual dimorphism in size (Kanazawa 2005a), although not of the relationship between polygyny and sexual dimorphism. The application of the generalized Trivers-Willard hypothesis (see "Boy or Girl? What Influences the Sex of Your Child?" in chapter 5) suggests that tall and heavy parents (both mothers and fathers) are more likely to have sons, and short and light parents are more likely to have daughters, because large body size is more adaptive for men than for women. Available empirical evidence supports this prediction of the generalized Trivers-Willard hypothesis. Body size (height and weight) are substantially heritable. Sexual dimorphism in size may therefore have evolved through this mechanism *in addition to* the effect of polygyny on the age of puberty.
22 Kirkpatrick (1987); Small (1993); Trivers (1972).
23 Kanazawa and Still (1999).
24 Shaw (1957, p. 254).
25 Dawkins (1986).
[26] Davies (1989); Orians (1969); Searcy and Yasukawa (1989); Verner (1964); Verner and Willson (1966). Borgerhoff Mulder (1990) is an earlier application of the polygyny threshold model to human society.
27 Lenski (1966, pp. 308–18).
28 Kanazawa and Still (2001).
29 Betzig (1986).
30 Kanazawa (2003a); Pérusse (1993, 1994).
31 Kanazawa and Still (1999).
32 Katzev, Warner, and Acock (1994); Morgan, Lye, and Condran (1988).
33 Draper and Harpending (1982).
34 Thornhill (1976).
35 Sozou and Seymour (2005).
36 Gangestad and Simpson (2000).
37 Draper and Harpending (1982).
38 Trivers (1972).
39 Gangestad and Thornhill (1997).

40 Rhodes, Simmons, and Peters (2005).
41 Gangestad and Simpson (2000, p. 583).

Chapter 5

1 Daly and Wilson (1985).
2 Daly and Wilson (1999).
3 Trivers and Willard (1973).
4 Betzig (1986).
5 Betzig and Weber (1995).
6 Cronk (1989).
7 Voland (1984).
8 Moore (1990, pp. 326–7, figures 1–2).
9 Mueller (1993).
10 Kanazawa (2006c).
11 Ellis and Bonin (2002); Freese and Powell (1999); Keller, Nesse, and Hofferth (2001).
12 Cronk (1991); Gaulin and Robbins (1991); Kanazawa (2001d); Trivers (2002, pp. 120–2).
13 Kanazawa (2005a, 2006b, 2007); Kanazawa and Vandermassen (2005).
14 Baron-Cohen (1999, 2002, 2003); Baron-Cohen and Hammer (1997); Baron-Cohen, Lutchmaya, and Knickmeyer (2004).
15 Kanazawa and Vandermassen (2005).
[16] The regression equations in Kanazawa and Vandermassen (2005, p. 595, table 1) include control variables for the respondent's education and income (to control for the effect of parental social class predicted in the original Trivers-Willard hypothesis), as well as age, age at first marriage, race, and current marital status. Then, controlling for the number of children of the opposite sex, having a systemizing occupation increases the number of sons by 0.3498 ($p < 0.01$), and having an empathizing occupation increases the number of daughters by 0.3981 ($p < 0.01$).

Let $S_O$ = the mean number of sons among the general population, $D_O$ = the mean number of daughters among the general population, $S_E$ = the mean number of sons among engineers and other systemizers, and $D_N$ = the mean number of daughters among nurses and other empathizers.

For our computation of $S_E$, assume $D_O = D_E = 1$. Then, $S_O = 1.0500$, and $S_E = 1.0500 + .3498 = 1.3998$. For our computation of $D_N$, assume $S_O = S_E = 1$. Then, $D_O = 0.9524$, and $D_N = 0.9524 + 0.3981 = 1.3505$. We thank Jouni Kuha for help with these computations.

17 Kanazawa (2005a).
18 Chagnon (1997).
19 de Waal (1982).
20 Kanazawa (2006b).
21 Kanazawa (2007).
22 Kanazawa (2007); Takahashi et al. (2006).
23 Christenfeld and Hill (1995).
24 McLain et al. (2000).
25 Brédart and French (1999); Bressan and Grassi (2004).
26 Daly and Wilson (1982); McLain et al. (2000); Regalski and Gaulin (1993).
27 McLain et al. (2000).
28 Kanazawa and Still (2000a, p. 25, appendix).
29 Liss (1987, p. 781).
30 US Bureau of the Census (1995, p. 7, table B).
31 Bellis et al. (2005).
32 Kanazawa and Still (2000a).
33 Daly and Wilson (1988, pp. 62–3).
34 Campbell (1988); Fischer and Oliker (1983); Marsden (1987).
35 Smith-Lovin and McPherson (1993); Munch, McPherson, and Smith-Lovin (1997).
36 Smith-Lovin and McPherson (1993, pp. 234–5).
37 Kanazawa (2001b).
38 Draper and Harpending (1982, 1988); Ellis et al. (1999).
39 Ellis et al. (2003); Quinlan (2003).
40 Ellis (2004, pp. 922–4).
41 Kaprio et al. (1995); Rowe (2002).
42 Ellis (2004).
43 Draper and Harpending (1982, 1988).
44 Quinlan (2003).
45 Bailey et al. (2000); Kanazawa (2001c).
46 Kanazawa (2001c).
47 Herman-Giddens et al. (1997); Lemonick (2000).
48 Ellis (2002).

### Chapter 6

1 Daly and Wilson (1988).
2 Brown (1991).
3 Pinker (2002, pp. 435–9, appendix).

4 Kanazawa (2006c).
5 International Criminal Police Organization (various years).
6 Daly and Wilson (1988, pp. 137–61).
7 Daly and Wilson (1988, pp. 123–36).
8 Wolfgang (1958).
9 Buss (1994, pp. 19–48).
10 Thornhill and Palmer (2000); Thornhill and Thornhill (1983).
11 Ellis (1998).
12 Campbell (1995, 1999).
13 Campbell (2002).
14 Campbell (1999, p. 210).
15 Browne (2002); Furchtgott-Roth and Stolba (1999); Kanazawa (2005b).
16 Greenberg (1985); Hirschi and Gottfredson (1985); Steffensmeier et al. (1989).
17 Blumstein (1995); Campbell (1995); Daly and Wilson (1990).
18 Miller (1999, p. 87; emphases added).
19 Kanazawa (2003c).
20 Kanazawa (2000); Miller (1999).
21 Kanazawa (2003c).
22 Kanazawa (2003b, 2003c); Kanazawa and Still (2000c); Miller (2000).
23 Kanazawa (2003c).
24 Trivers (1972).
25 Blumstein and Schwartz (1983, pp. 195–8); Laumann et al. (1994, pp. 315–6, table 8.4).
26 Kanazawa (2000, 2003c); Miller (1998, 1999, 2000).
27 Laub, Nagin and Sampson (1998); Sampson and Laub (1993).
28 Hirschi (1969).
29 Hirschi (1969).
30 Kanazawa (2000).
31 Hargens, McCann, and Reskin (1978).
32 Gould and Lewontin (1979).
33 Ketelaar and Ellis (2000); Kurzban and Haselton (2006).
34 Daly and Wilson (1996).
35 Wilson, Daly, and Wright (1993); Wilson, Johnson, and Daly (1995).
36 Wilson, Daly, and Wright (1993, p. 275, table 4).
37 Wilson, Daly, and Wright (1993, p. 276, table 5).
38 Kanazawa and Still (2000c, pp. 444–6).

39 Buss (1988); Buss and Shackelford (1997); Peters, Shackelford, and Buss (2002).
40 Wilson, Daly, and Wright (1993).

## Chapter 7

[1] The only other American President ever to be impeached was Andrew Johnson, who was elected Vice President and became President after Abraham Lincoln's assassination in 1865. As a result of the Watergate scandal, the House of Representatives began impeachment hearings against Richard M. Nixon in 1974. However, Nixon resigned in August 1974 before the full House had a chance to vote on the impeachment, the first (and so far the only) US President ever to resign.
2 Betzig (1982, 1986, 1993, 2002).
3 Betzig (1992, 1995).
4 Kanazawa (2004c).
5 Blau and Kahn (1992); Mueller, Kuruvilla, and Iverson (1994); Rosenfeld and Kalleberg (1990); Sørensen and Trappe (1995).
6 Blau and Kahn (2000).
7 Marini (1989).
8 England (1992).
9 Browne (1995, 1998, 2002).
10 Campbell (1999, 2002).
11 Kanazawa (2005b, p. 276, table 1).
12 Kanazawa (2005b, p. 284).
13 Moir and Jessel (1989, p. 167).
14 Moir and Jessel (1989, p. 159).
15 Kanazawa (2005b).
16 Eitzen (1985, p. 378); Furchtgott-Roth and Stolba (1999, p. 11).
17 Eitzen (1988, p. 385).
18 Eitzen and Zinn (1991, p. 324).
19 "Clinton Seeks More Money to Reduce Gap in Wages." *New York Times*, January 31, 1999.
20 Furchtgott-Roth and Stolba (1999).
21 Kanazawa (2005b).
22 Browne (2002).
23 Baron-Cohen (1999, 2002, 2003); Baron-Cohen and Hammer (1997); Baron-Cohen, Lutchmaya, and Knickmeyer (2004).

24 Baron-Cohen (2003, p. 3).
25 Baron-Cohen (2003, p. 63).
26 Baron-Cohen (2003, p. 2).
27 Baron-Cohen (2003, pp. 23–4).
28 Baron-Cohen (2003, p. 24).
29 Baron-Cohen (2003, p. 60, figure 5; p. 85, figure 7).
30 Baron-Cohen et al. (1997, 1998).
31 Browne (1997, 2002, pp. 191–214; 2006).
32 Franke (1995); Paludi (1996); Tangri, Burt, and Johnson (1982).
33 Browne (1997).
34 Clarke and Hatfield (1989).
35 Buss and Schmitt (1993).
36 Avner (1994); Bravo and Cassedy (1992).
37 Browne (2002, p. 202).
38 Muehlenhard and Hollabaugh (1988).
39 Muehlenhard and McCoy (1991).
40 Mealey (1992, p. 397).
41 Browne (1997, p. 75).

## Chapter 8

1 Kurzban, Tooby, and Cosmides (2001).

[2] Before we can explore the evolutionary origins of religion, we must first define our terms. The term *religion*, both in academic and general writing, tends to refer to three related yet separate things: *religious beliefs* (intraindividual cognitive processes inside the brain); *religious practices* (individual and interindividual social behavior, such as rituals); and *religious organization* (supraindividual collectivities, such as churches, synagogues, and other denominations). Psychologists mostly study religious beliefs (Allport 1950; James 1902), anthropologists usually focus on religious beliefs and practices (Durkheim 1915/1965; Evans-Pritchard 1956), and sociologists and economists tend to concentrate on religious practices and organizations (Greeley 1972; Iannaccone 1994).

In this section, we focus exclusively on the evolutionary psychological origins of *religious beliefs*. For this reason, we exclude from our discussion David Sloan Wilson's excellent book *Darwin's Cathedral: Evolution, Religion, and the Nature of Society* (2002), because it is mostly about religious organization and how different religious groups and societies evolved over history.

3 Brown (1991).

4 Bouchard et al. (1999); Koenig et al. (2005).

5 Alper (2001); Hamer (2004); Newberg, D'Aquili, and Rause (2002).

[6] There have been a few studies that conclude that religious and spiritual people live longer than nonbelievers (Hall 2006; Miller and Thoresen 2003; McCullough et al. 2000). However, no one has yet either specified the *proximate* biochemical mechanism of *how* religiosity increases longevity or explained the *ultimate* reason *why* it does. These studies are also of limited credibility, since they are either funded by the John Templeton Foundation (Miller and Thoresen 2003; McCullough et al. 2000) or conducted by an Episcopal priest (Hall 2006).

7 Kirkpatrick (2005, pp. 214–39).

8 Atran (2002); Boyer (2001); Guthrie (1993); Kirkpatrick (2005).

[9] Once again, if you are familiar with elementary statistics, you recognize the false-positive and false-negative errors as "Type I" and "Type II" errors.

10 Haselton and Nettle (2006).

11 Guthrie (1993).

12 Atran (2002).

[13] In other words, according to these theorists, religion is the result of humans attributing an intention to, and thereby employing theory of mind for, inanimate physical objects. McNamara (2001) suggests that autistics, who lack the theory of mind module, may thus be less likely to be religious than nonautistics. To the best of our knowledge, no one has compared the levels of religiosity among autistics and nonautistics.

14 Haselton (2003); Haselton and Buss (2000).

[15] Haselton and Nettle (2006). The fact that Kirkpatrick's theory of the evolution of religion has much to do with Haselton's error management theory may not be entirely coincidental. Haselton was once a student of Kirkpatrick's, although it was Haselton who originally introduced Kirkpatrick to evolutionary psychology (Kirkpatrick 2005, pp. x–xi).

[16] In Bangladesh, very slightly more men believe in God (98.7% vs. 98.5%) and identify themselves as religious (84.1% vs. 83.7%) than women. In the Dominican Republic, more men believe in God than women (95.2% vs. 90.9%), but the sample size is small (411 respondents). In Montenegro, more men believe in God (66.3% vs. 63.9%) and identify themselves as religious (50.9% vs. 47.9%) than women, but the sample size is even smaller (209 for the first question, 225 for the second).

17 Miller and Stark (2002).

18 Mol (1985); Suziedalis and Potvin (1981).

19 Glock, Ringer, and Babbie (1967); Walter and Davie (1998).
20 Azzi and Ehrenberg (1975); Iannaccone (1990); Luckmann (1967); Martin (1967).
21 Miller and Stark (2002).
22 de Vaus and McAllister (1987); Steggarda (1993).
23 Cornwall (1988); de Vaus (1984); Stark (1992).
24 Haselton and Nettle (2006).
25 Kanazawa and Still (2000c).
26 Campbell (1995, 1999, 2002).
27 Miller and Hoffmann (1995); Miller and Stark (2002); Stark (2002); Sherkat (2002).
28 Miller (2000).
29 Gambetta (2005, pp. 259–63).
30 Kanazawa and Still (2000c).
[31] The top 20 most polygynous nations according to these scores are: 1. Anguilla, 1. Antigua and Barbuda, 1. Bahamas, 1. Barbados, 1. Equitorial Guinea, 1. Gabon, 1. Haiti, 1. Lesotho, 1. St. Vincent/Grenadines, 1. Swaziland (all of which have the maximum polygyny score of 3.000), 11. Morocco (2.9700), 12. Liberia (2.9000), 13. Nigeria (2.8175), 14. Congo (former Zaire) (2.8095), 15. Sierra Leone (2.8000), 16. Chad, 16. Nicaragua (both 2.7500), 18. Niger (2.7250), 19. Togo (2.6667), and 20. Mozambique (2.6664). Only Morocco and Nicaragua are outside of sub-Saharan Africa and the Caribbean.
32 Atran (2003); Berrebi (2003).
33 O'Hanlon and Campbell (2007).
34 Kalyvas (2005, pp. 96–7).
35 Coogan (1995, pp. 513–21).
36 Atran (2003, p. 1538); Friedman (2002, pp. 144–5).
37 Krueger and Maleckova (2003, p. 129).
38 Friedman (2002, pp. 13–4, 19–20).
39 Hechter (2000).
40 Coleman (1988).
41 Olson (1965).
42 Kanazawa (2001a).
43 Whitmeyer (1997).
44 Dawkins (1976).
45 Pinker (2002).
46 Kurzban, Tooby, and Cosmides (2001).

47 Miller (2000).
48 Townsend and Levy (1990).
49 Dunbar, Duncan, and Marriott (1997).
50 Kanazawa (2000, 2003c); Miller (1998, 1999).
51 Lycett and Dunbar (2000).
52 Low (1979).
53 Cunningham et al. (1995); Jones (1996); Jones and Hill (1993); Maret and Harling (1985); Morse and Gruzen (1976); Thakerar and Iwawaki (1979).
54 Langlois et al. (1987); Samuels and Ewy (1985).
55 Buss (1999, p. 135).
56 Kanazawa (2000, 2003c).
57 Dugatkin (1998).
58 Grant and Green (1998).
59 Höglund et al. (1995).
60 Galef and White (1998).
61 Dugatkin (2000).
[62] While an appealing idea, the only experimental test of this "wedding-ring effect" has not been supportive (Uller and Johansson 2003), so this must still be treated as an interesting but speculative idea.
63 Kanazawa and Frerichs (2001).
64 Kanazawa and Frerichs (2001, p. 327, table 2).

### Conclusion

1 Miller (2000, pp. 217–9).
2 Hamer et al. (1993).
3 Hamer and Copeland (1994, pp. 183–4).
4 Trivers (personal communication).
5 Camperio-Ciani, Corna, and Capiluppi (2004).
6 Hamer and Copeland (1994, pp. 182–3); Miller (2000, pp. 217–9).
[7] Sulloway has had a truly maverick academic career. Having received a PhD in History of Science at Harvard and having been mentored by the great evolutionary biologist Ernst Mayr, Sulloway has never held a regular academic appointment, and has instead supported himself and his scientific research entirely through research grants and fellowships, including a MacArthur Prize Fellowship (i.e., "the genius award"). He has written on the history of science, psychology, and evolutionary biology, and has conducted research at Harvard, MIT, the Center for Advanced Study in the Behavioral Sciences at Stanford, and the University of California, Berkeley, where he is

currently Visiting Scholar and Professor (Sulloway, personal communication).

[8] In 1960, Judith Rich Harris was a graduate student in psychology at Harvard. After receiving her master's degree, she was dismissed from the program by the then acting department chair, George A. Miller, who thought Harris was not smart enough to earn a PhD. Thirty-five years later, while supporting herself by writing psychology textbooks, Harris worked on her group socialization theory of development and published it in the prestigious academic journal *Psychological Review.* In 1997, her article won an award from the American Psychological Association, the George A. Miller Award for an Outstanding Recent Article in General Psychology (Harris 1998, pp. xi–xviii).

9 Sulloway (1996).

10 Harris (1995, 1998).

11 Rowe (1994).

12 Harris (1998, pp. 365–78); Sulloway (2000).

13 Kohler, Rodgers, and Christensen (1999); Rodgers, Kohler, Kyvik, and Christensen (2001); Rodgers et al. (2001).

14 Daly and Wilson (1985).

[15] Daly and Wilson (1995). Probably the most common cause of mothers killing their biological children is mental illness. With its emphasis on universal human nature, however, evolutionary psychology is ill equipped to explain behavior caused by mental illness and other "abnormality."

16 Daly and Wilson (1988, pp. 37–93).

17 Shields and Shields (1983); Thornhill and Palmer (2000).

18 Yamaguchi and Ferguson (1995).

19 Saad and Peng (2006).

20 Kanazawa (2006c).

21 Buss (1989).

22 Schmitt (2003, 2004).

23 Wright (1994, p. 84).

24 Low, personal communication.

# References

Abbey, Antonia. 1982. "Sex Differences in Attributions for Friendly Behavior: Do Males Misperceive Females' Friendliness?" *Journal of Personality and Social Psychology.* 42: 830–8.

Abdollahi, Panteha and Traci Mann. 2001. "Eating Disorder Symptoms and Body Image Concern in Iran: Comparisons Between Iranian Women in Iran and in America." *International Journal of Eating Disorders.* 30: 259–68.

Al-Eisa, Einas, David Egan, and Richard Wassersub. 2004. "Fluctuating Asymmetry and Low Back Pain." *Evolution and Human Behavior.* 25: 31–7.

Alexander, Gerianne M. and Melissa Hines. 2002. "Sex Differences in Response to Children's Toys in Nonhuman Primates (*Cercopithecus aethiops sabaeus*)." *Evolution and Human Behavior.* 23: 467–79.

Alexander, Richard D., John L. Hoogland, Richard D. Howard, Katharine M. Noonan, and Paul W. Sherman. 1979. "Sexual Dimorphisms and Breeding Systems in Pinnipeds, Ungulates, Primates and Humans." Pp. 402–35 in *Evolutionary Biology and Human Social Behavior: An Anthropological Perspective,* edited by Napoleon A. Chagnon and William Irons. North Scituate: Duxbury Press.

Allport, Gordon W. 1950. *The Individual and His Religion.* New York: Macmillan.

Alper, Matthew. 2001. *The "God" Part of the Brain: A Scientific Interpretation of Human Spirituality and God.* New York: Rogue Press.

Atran, Scott. 2002. *In Gods We Trust: The Evolutionary Landscape of Religion.* Oxford: Oxford University Press.

Atran, Scott. 2003. "Genesis of Suicide Terrorism." *Science.* 299: 1534–39.

Avner, Judith I. 1994. "Sexual Harassment: Building a Consensus for Change." *Kansas Journal of Law and Public Policy.* 3: 57–76.

Azzi, Corry and Ronald Ehrenberg. 1975. "Household Allocation of Time and Church Attendance." *Journal of Political Economy.* 83: 27–56.

Bailey, J. Michael, Katherine M. Kirk, Gu Zhu, Michael P. Dunne, and Nicholas G. Martin. 2000. "Do Individual Differences in Sociosexuality Represent Genetic or Environmentally Contingent Strategies? Evidence from the

Australian Twin Registry." *Journal of Personality and Social Psychology*. 78: 537–45.

Bailit, H. L., P. L. Workman, J. D. Niswander, and J. C. Maclean. 1970. "Dental Asymmetry as an Indicator of Genetic and Environmental Conditions in Human Populations." *Human Biology*. 42: 626–38.

Baker, R. Robin and Mark A. Bellis. 1995. *Human Sperm Competition: Copulation, Masturbation and Infidelity*. London: Chapman and Hall.

Barash, David P. 1982. *Sociobiology and Behavior*, Second Edition. New York: Elsevier.

Barash, David P. and Judith Eve Lipton. 2001. *The Myth of Monogamy: Fidelity and Infidelity in Animals and People*. New York: W. H. Freeman.

Barkow, Jerome H. (Editor.) 2006. *Missing the Revolution: Darwinism for Social Scientists*. Oxford: Oxford University Press.

Barkow, Jerome H., Leda Cosmides, and John Tooby. (Editors.) 1992. *The Adapted Mind: Evolutionary Psychology and the Generation of Culture*. New York: Oxford University Press.

Baron-Cohen, Simon. 1999. "The Extreme Male Brain Theory of Autism." Pp. 401–29 in *Neurodevelopmental Disorders*, edited by Helen Tager-Flusberg. Cambridge: MIT Press.

Baron-Cohen, Simon. 2002. "The Extreme Male Brain Theory of Autism." *Trends in Cognitive Science*. 6: 248–54.

Baron-Cohen, Simon. 2003. *The Essential Difference*. London: Penguin.

Baron-Cohen, Simon, Patrick Bolton, Sally Wheelwright, Victoria Scahill, Liz Short, Genevieve Mead, and Alex Smith. 1998. "Autism Occurs More Often in Families of Physicists, Engineers, and Mathematicians." *Autism*. 2: 296–301.

Baron-Cohen, Simon and Jessica Hammer. 1997. "Is Autism an Extreme Form of the Male Brain?" *Advances in Infancy Research*. 11: 193–217.

Baron-Cohen, Simon, Svetlana Lutchmaya, and Rebecca Knickmeyer. 2004. *Prenatal Testosterone in Mind: Amniotic Fluid Studies*. Cambridge: MIT Press.

Baron-Cohen, Simon, Sally Wheelwright, Carol Stott, Patrick Bolton, and Ian Goodyer. 1997. "Is There a Link Between Engineering and Autism?" *Autism*. 1: 101–9.

Bellis, Mark A. and Robin Baker. 1990. "Do Females Promote Sperm Competition? Data for Humans." *Animal Behaviour*. 40: 997–9.

Bellis, Mark A., Karen Hughes, Sara Hughes, and John R. Ashton. 2005. "Measuring Paternal Discrepancy and Its Public Health Consequences." *Journal of Epidemiology and Community Health*. 59: 749–54.

Bernstein, Ira H., Tsai-Ding Lin, and Pamela McClellan. 1982. "Cross- vs. Within-Racial Judgments of Attractiveness." *Perception and Psychophysics.* 32: 495–503.

Berrebi, Claude. 2003. "Evidence about the Link Between Education, Poverty and Terrorism among Palestinians." Princeton University Industrial Relations Sections Working Paper #477.

Betzig, Laura. 1982. "Despotism and Differential Reproduction: A Cross-Cultural Correlation of Conflict Asymmetry, Hierarchy, and Degree of Polygyny." *Ethology and Sociobiology.* 3: 209–21.

Betzig, Laura L. 1986. *Despotism and Differential Reproduction: A Darwinian View of History.* New York: Aldine.

Betzig, Laura. 1989. "Causes of Conjugal Dissolution: A Cross-Cultural Study." *Current Anthropology.* 30: 654–76.

Betzig, Laura. 1992. "Roman Polygyny." *Ethology and Sociobiology.* 13: 309–49.

Betzig, Laura. 1993. "Sex, Succession, and Stratification in the First Six Civilizations: How Powerful Men Reproduced, Passed Power on to Their Sons, and Used Power to Defend Their Wealth, Women, and Children." Pp. 37–74 in *Social Stratification and Socioeconomic Inequality.* Volume 1: A Comparative Biosocial Analysis, edited by Lee Ellis. Westport: Praeger.

Betzig, Laura. 1995. "Medieval Monogamy." *Journal of Family History.* 20: 181–216.

Betzig, Laura. 1997a. "People Are Animals." Pp. 1–17 in *Human Nature: A Critical Reader,* edited by Laura Betzig. New York: Oxford University Press.

Betzig, Laura. (Editor.) 1997b. *Human Nature: A Critical Reader.* New York: Oxford University Press.

Betzig, Laura. 2002. "British Polygyny." Pp. 30–97 in *Human Biology and History,* edited by Malcolm Smith. London: Taylor and Francis.

Betzig, Laura and Samantha Weber. 1995. "Presidents Preferred Sons." *Politics and the Life Sciences.* 14: 61–4.

Birkhead, T. R. and Anders Pape Møller. (Editors.) 1991. *Sperm Competition in Birds: Evolutionary Causes and Consequences.* London: Academic Press.

Biro, Frank M., Robert P. McMahon, Ruth Striegel-Moore, Patricia B. Crawford, Eva Obarzanek, John A. Morrison, Bruce A. Barton, and Frank Falkner. 2001. "Impact of Timing of Pubertal Maturation on Growth in Black and White Female Adolescents: The National Heart, Lung, and Blood Institute Growth and Health Study." *Journal of Pediatrics.* 138: 636–43.

Blau, Francine D. and Lawrence M. Kahn. 1992. "The Gender Earnings Gap: Learning from International Comparisons." *American Economic Review.* 82 (May): 533–8.

Blau, Francine D. and Lawrence M. Kahn. 2000. "Gender Differences in Pay." *Journal of Economic Perspectives.* 14 (4): 75–99.

Bloch, Konrad. 1994. *Blondes in Venetian Paintings, the Nine-Banded Armadillo, and Other Essays in Biochemistry.* New Haven: Yale University Press.

Blum, Deborah. 1997. *Sex on the Brain: The Biological Differences Between Men and Women.* New York: Penguin.

Blumstein, Alfred. 1995. "Youth Violence, Guns, and the Illicit-Drug Industry." *Journal of Criminal Law and Criminology.* 86: 10–36.

Blumstein, Philip, and Pepper Schwartz. 1983. *American Couples: Money, Work, Sex.* Pocket Books.

Borgerhoff Mulder, Monique. 1990. "Kipsigis Women's Preference for Wealthy Men: Evidence for Female Choice in Mammals." *Behavioral Ecology and Sociobiology.* 27: 255–64.

Bouchard, Jr., Thomas J., Matt McGue, David Lykken, and Auke Tellegen. 1999. "Intrinsic and Extrinsic Religiousness: Genetic and Environmental Influences and Personality Correlates." *Twin Research.* 2: 88–98.

Bowlby, John. 1969. *Attachment and Loss.* Volume 1: Attachment. New York: Basic.

Boyer, Pascal. 2001. *Religion Explained: The Evolutionary Origins of Religious Thought.* New York: Basic.

Bravo, Ellen and Ellen Cassedy. 1992. *The 9 to 5 Guide to Combating Sexual Harassment.* New York: Wiley.

Brédart, Serge and Robert M. French. 1999. "Do Babies Resemble Their Fathers More Than Their Mothers? A Failure to Replicate Christenfeld and Hill (1995)." *Evolution and Human Behavior.* 20: 129–35.

Bressan, Paola and Massimo Grassi. 2004. "Parental Resemblance in 1-Year-Olds and the Gaussian Curve." *Evolution and Human Behavior.* 25: 133–41.

Brown, Donald E. 1991. *Human Universals.* New York: McGraw-Hill.

Browne, Kingsley R. 1995. "Sex and Temperament in Modern Society: A Darwinian View of the Glass Ceiling and the Gender Gap." *Arizona Law Review.* 37: 971–1106.

Browne, Kingsley R. 1997. "An Evolutionary Perspective on Sexual Harassment: Seeking Roots in Biology Rather Than Ideology." *Journal of Contemporary Legal Issues.* 8: 5–77.

Browne, Kingsley. 1998. *Divided Labours: An Evolutionary View of Women at Work.* London: Weidenfeld and Nicolson.

Browne, Kingsley R. 2002. *Biology at Work: Rethinking Sexual Equality.* New Brunswick: Rutgers University Press.

Browne, Kingsley R. 2006. "Sex, Power, and Dominance: The Evolutionary Psychology of Sexual Harassment." *Managerial and Decision Economics.* 27: 145–58.

Buss, David M. 1988. "From Vigilance to Violence: Tactics of Mate Retention." *Ethology and Sociobiology.* 9: 291–317.

Buss, David M. 1989. "Sex Differences in Human Mate Preferences: Evolutionary Hypotheses Tested in 37 Cultures." *Behavioral and Brain Sciences.* 12: 1–49.

Buss, David M. 1994. *The Evolution of Desire: Strategies of Human Mating.* New York: Basic.

Buss, David M. 1995. "Evolutionary Psychology: A New Paradigm for Psychological Science." *Psychological Inquiry.* 6: 1–30.

Buss, David M. 1999. *Evolutionary Psychology: The New Science of the Mind.* Boston: Allyn and Bacon.

Buss, David M. 2000. *The Dangerous Passion: Why Jealousy Is as Necessary as Love and Sex.* New York: Free Press.

Buss, David M., Randy J. Larsen, and Drew Westen. 1992. "Sex Differences in Jealousy: Evolution, Physiology, and Psychology." *Psychological Science.* 3: 251–5.

Buss, David M. and David P. Schmitt. 1993. "Sexual Strategies Theory: An Evolutionary Perspective on Human Mating." *Psychological Review.* 100: 204–32.

Buss, David M. and Todd K. Shackelford. 1997. "From Vigilance to Violence: Mate Retention Tactics in Married Couples." *Journal of Personality and Social Psychology.* 72: 346–61.

Buss, David M., Todd K. Shackelford, Lee A. Kirkpatrick, Jae C. Choe, Mariko Hasegawa, Toshikazu Hasegawa, and Kevin Bennett. 1999. "Jealousy and the Nature of Beliefs about Infidelity: Tests of Competing Hypotheses about Sex Differences in the United States, Korea, and Japan." *Personal Relationships.* 6: 125–50.

Buunk, Bram and Ralph B. Hupka. 1987. "Cross-Cultural Differences in the Elicitation of Sexual Jealousy." *Journal of Sex Research.* 23: 12–22.

Calden, George, Richard M. Lundy, and Richard J. Schlafer. 1959. "Sex Differences in Body Concepts." *Journal of Consulting Psychology.* 23: 378.

Campbell, Anne. 1995. "A Few Good Men: Evolutionary Psychology and Female Adolescent Aggression." *Ethology and Sociobiology.* 16: 99–123.

Campbell, Anne. 1999. "Staying Alive: Evolution, Culture, and Women's Intrasexual Aggression." *Behavior and Brain Sciences.* 22: 203–52.

Campbell, Anne. 2002. *A Mind of Her Own: The Evolutionary Psychology of Women.* Oxford: Oxford University Press.

Campbell, Karen E. 1988. "Gender Differences in Job-Related Networks." *Work and Occupations.* 15: 179–200.

Camperio-Ciani, Andrea, Francesca Corna, and Claudio Capiluppi. 2004. "Evidence for Maternally Inherited Factors Favouring Male Homosexuality and Promoting Female Fecundity." *Proceedings of the Royal Society of London, Series B.* 271: 2217–21.

Carroll, Joseph. 1999. "The Deep Structure of Literary Representations." *Evolution and Human Behavior.* 20: 159–73.

Cartwright, John. 2000. *Evolution and Human Behavior: Darwinian Perspectives on Human Nature.* Cambridge: MIT Press.

Cerda-Flores, Ricardo M., Sara A. Barton, Luisa F. Marty-Gonzalez, Fernando Rivas, and Ranajit Chakraborty. 1999. "Estimation of Nonpaternity in the Mexican Population of Nuevo Leon: A Validation Study with Blood Group Markers." *American Journal of Physical Anthropology.* 109: 281–93.

Chagnon, Napoleon A. 1968. *Yanomamö: The Fierce People.* New York: Holt, Rinehart, and Winston.

Chagnon, Napoleon A. 1997. *Yanomamö*, Fifth Edition. Fort Worth: Harcourt Brace.

Chisholm, James S. and Victoria K. Burbank. 1991. "Monogamy and Polygyny in Southeast Arnhem Land: Male Coercion and Female Choice." *Ethology and Sociobiology.* 12: 291–313.

Chomsky, Noam. 1957. *Syntactic Structures.* The Hague: Mouton.

Christenfeld, Nicholas J. S. and Emily A. Hill. 1995. "Whose Baby Are You?" *Nature.* 378: 669.

Clarke, Russell D. and Elaine Hatfield. 1989. "Gender Differences in Receptivity to Sexual Offers." *Journal of Psychology and Human Sexuality.* 2: 39–55.

Clutton-Brock, T. H. and A.C.J. Vincent. 1991. "Sexual Selection and the Potential Reproductive Rates of Males and Females." *Nature.* 351: 58–60.

Coleman, James S. 1988. "Freeriders and Zealots: The Role of Social Networks." *Sociological Theory.* 6: 52–57.

Connellan, Jennifer, Simon Baron-Cohen, Sally Wheelwright, Anna Batki, and Jag Ahluwalia. 2000. "Sex Differences in Human Neonatal Social Perception." *Infant Behavior and Development.* 23: 113–8.

Coogan, Tim Pat. 1995. *The I.R.A.*, Fourth Edition. London: HarperCollins.

Cornwall, Marie. 1988. "The Influence of Three Agents of Religious Socialization." Pp. 207–31 in *The Religion and Family Connection*, edited by Darwin Thomas. Provo: Brigham Young University Religious Studies Center.

Cornwell, R. Elisabeth, Kristi M. Hetterscheidt, Craig T. Palmer, and Hasker

P. Davis. 2001. The Status of Sociobiology/Evolutionary Psychology in Sociology. Paper presented at the Annual Meetings of the Human Evolution and Behavior Society, London, June 13–17.

Cornwell, R. Elisabeth, Craig T. Palmer, and Hasker P. Davis. 2000. Sociobiology and Evolutionary Psychology: A 25-Year Retrospective on Change and Treatment in Psychology. Paper presented at the Annual Meetings of the Human Evolution and Behavior Society, Amherst, June 7–11.

Crawford, Charles B. 1993. "The Future of Sociobiology: Counting Babies or Proximate Mechanisms." *Trends in Ecology and Evolution.* 8: 183–6.

Crawford, Charles B., Brenda E. Salter, and Kerry L. Jang. 1989. "Human Grief: Is Its Intensity Related to the Reproductive Value of the Deceased?" *Ethology and Sociobiology.* 10: 297–307.

Cronin, Helena. 1991. *The Ant and the Peacock: Altruism and Sexual Selection from Darwin to Today.* Cambridge: Cambridge University Press.

Cronk, Lee. 1989. "Low Socioeconomic Status and Female-Based Parental Investment: The Mukogodo Example." *American Anthropologist.* 91: 414–29.

Cronk, Lee. 1991. "Preferential Parental Investment in Daughters over Sons." *Human Nature.* 2: 387–417.

Cross, John F. and Jane Cross. 1971. "Age, Sex, Race, and the Perception of Facial Beauty." *Developmental Psychology.* 5: 433–9.

Cunningham, Michael R., Perri B. Druen, and Anita P. Barbee. 1997. "Angels, Mentors, and Friends: Trade-Offs among Evolutionary, Social, and Individual Variables in Physical Appearance." Pp. 109–40 in *Evolutionary Social Psychology,* edited by Jeffry A. Simpson and Douglas T. Kenrick. Mahwah: Lawrence Erlbaum.

Cunningham, Michael R., Alan R. Roberts, Anita P. Barbee, Perri B. Druen, and Cheng-Huan Wu. 1995. "'Their Ideas of Beauty Are, on the Whole, the Same as Ours': Consistency and Variability in the Cross-Cultural Perception of Female Physical Attractiveness." *Journal of Personality and Social Psychology.* 68: 261–79.

Daly, Martin and Margo I. Wilson. 1982. "Whom Are Newborn Babies Said to Resemble?" *Ethology and Sociobiology.* 3: 69–78.

Daly, Martin and Margo Wilson. 1985. "Child Abuse and Other Risks of Not Living with Both Parents." *Ethology and Sociobiology.* 6: 197–210.

Daly, Martin and Margo Wilson. 1988. *Homicide.* New York: De Gruyter.

Daly, Martin and Margo Wilson. 1990. "Killing the Competition: Female/Female and Male/Male Homicide." *Human Nature.* 1: 81–107.

Daly, Martin and Margo Wilson. 1995. "Discriminative Parental Solicitude and the Relevance of Evolutionary Models to the Analysis of Motivational Systems." Pp. 1269–86 in *The Cognitive Neurosciences*, edited by Michael S. Gazzaniga. Cambridge: MIT Press.

Daly, Martin and Margo Wilson. 1996. "Homicidal Tendencies." *Demos*. 10: 39–45.

Daly, Martin and Margo Wilson. 1999. *The Truth about Cinderella: A Darwinian View of Parental Love*. New Haven: Yale University Press.

Daly, Martin, Margo Wilson, and Suzanne J. Weghorst. 1982. "Male Sexual Jealousy." *Ethology and Sociobiology*. 3: 11–27.

Darwin, Charles. 1859. *On the Origin of Species by Means of Natural Selection*. London: John Murray.

Darwin, Charles. 1871. *The Descent of Man, and Selection in Relation to Sex*. London: John Murray.

Davies, N. B. 1989. "Sexual Conflict and the Polygamy Threshold." *Animal Behaviour*. 38: 226–34.

Davis, Bernard. 1978. "Moralistic Fallacy." *Nature*. 272: 390.

Davis, Caroline, Colin M. Shapiro, Stuart Elliott, and Michelle Dionne. 1993. "Personality and Other Correlates of Dietary Restraint: An Age by Sex Comparison." *Personality and Individual Differences*. 14: 297–305.

Dawkins, Richard. 1976. *The Selfish Gene*. Oxford: Oxford University Press.

Dawkins, Richard. 1986. "Wealth, Polygyny, and Reproductive Success." *Behavioral and Brain Sciences*. 9: 190–1.

de Vaus, David A. 1984. "Workforce Participation and Sex Differences in Church Attendance." *Review of Religious Research*. 25: 247–58.

de Vaus, David A. and Ian McAllister. 1987. "Gender Differences in Religion: A Test of the Structural Location Theory." *American Sociological Review*. 52: 472–81.

de Waal, Frans B. M. 1982. *Chimpanzee Politics: Power and Sex among Apes*. London: Jonathan Cape.

DeSteno, David, Monica Y. Bartlett, Julia Braverman, and Peter Salovey. 2002. "Sex Differences in Jealousy: Evolutionary Mechanism or Artifact of Measurement?" *Journal of Personality and Social Psychology*. 83: 1103–16.

Draper, Patricia and Henry Harpending. 1982. "Father Absence and Reproductive Strategy: An Evolutionary Perspective." *Journal of Anthropological Research*. 38: 255–73.

Draper, Patricia and Henry Harpending. 1988. "A Sociobiological Perspective on the Development of Human Reproductive Strategies." Pp. 340–72 in *Sociobiological Perspectives on Human Development*, edited by Kevin B. MacDonald. New York: Springer-Verlag.

Dugatkin, Lee A. 1998. "Genes, Copying, and Female Mate Choice: Shifting Thresholds." *Behavioral Ecology.* 9: 323–7.

Dugatkin, Lee Alan. 2000. *The Imitation Factor: Evolution Beyond the Gene.* New York: Free Press.

Dunbar, R. I. M., N. D. C. Duncan, and Anna Marriott. 1997. "Human Conversational Behavior." *Human Nature.* 8: 231–46.

Durkheim, Emile. 1965 [1915]. *The Elementary Forms of the Religious Life.* New York: Free Press.

Eitzen, D. Stanley. 1985. *In Conflict and Order: Understanding Society*, Third Edition. Boston: Allyn and Bacon.

Eitzen, D. Stanley. 1988. *In Conflict and Order: Understanding Society*, Fourth Edition. Boston: Allyn and Bacon.

Eitzen, D. Stanley and Maxine Baca Zinn. 1991. *In Conflict and Order: Understanding Society*, Fifth Edition. Boston: Allyn and Bacon.

Ellis, Bruce J. 2002. "Of Fathers and Pheromones: Implications of Cohabitation for Daughters' Pubertal Timing." pp. 161–72 in *Just Living Together: Implications of Cohabitation on Families, Children, and Social Policy*, edited by Alan Booth and Ann C. Crouter. Mahwah: Lawrence Erlbaum.

Ellis, Bruce J. 2004. "Timing of Pubertal Maturation in Girls: An Integrated Life History Approach." *Psychological Bulletin.* 130: 920–58.

Ellis, Bruce J., John E. Bates, Kenneth A. Dodge, David M. Fergusson, L. John Horwood, Gregory S. Pettit, and Lianne Woodward. 2003. "Does Father Absence Place Daughters at Special Risk for Early Sexual Activity and Teenage Pregnancy?" *Child Development.* 74: 801–21.

Ellis, Bruce J. and David F. Bjorklund. 2005. "Preface." Pp. ix–xii in *Origins of the Social Mind: Evolutionary Psychology and Child Development*, edited by Bruce J. Ellis and David F. Bjorklund. New York: Guilford.

Ellis, Bruce J., Steven McFadyen-Ketchum, Kenneth A. Dodge, Gregory S. Pettit, and John E. Bates. 1999. "Quality of Early Family Relationships and Individual Differences in the Timing of Pubertal Maturation in Girls: A Longitudinal Test of an Evolutionary Model." *Journal of Personality and Social Psychology.* 77: 387–401.

Ellis, Bruce J. and Donald Symons. 1990. "Sex Differences in Fantasy: An Evolutionary Psychological Approach." *Journal of Sex Research.* 27: 527–57.

Ellis, Lee. 1996. "A Discipline in Peril: Sociology's Future Hinges on Curing Its Biophobia." *American Sociologist.* 27: 21–41.

Ellis, Lee. 1998. "Neo-Darwinian Theories of Violent Criminality and Antisocial Behavior: Photographic Evidence from Nonhuman Animals and a Review of the Literature." *Aggression and Violent Behavior.* 3: 61–110.

Ellis, Lee and Steven Bonin. 2002. "Social Status and the Second Sex Ratio: New Evidence on a Lingering Controversy." *Social Biology.* 49: 35–43.

Emlen, Stephen T. 1995. "Can Avian Biology Be Useful to the Social Sciences?" *Journal of Avian Biology.* 26: 273–6.

England, Paula. 1992. *Comparable Worth: Theories and Evidence.* New York: Aldine.

Etcoff, Nancy. 1999. *Survival of the Prettiest: The Science of Beauty.* New York: Doubleday.

Evans-Pritchard, E. E. 1956. *Nuer Religion.* New York: Oxford University Press.

Feinman, Saul and George W. Gill. 1978. "Sex Differences in Physical Attractiveness Preferences." *Journal of Social Psychology.* 105: 43–52.

Fischer, Claude S. and Stacy J. Oliker. 1983. "A Research Note on Friendship, Gender, and the Life Cycle." *Social Forces.* 62: 124–33.

Franke, Katherine M. 1995. "The Central Mistake of Sex Discrimination Law: The Disaggregation of Sex from Gender." *University of Pennsylvania Law Review.* 144: 1–99.

Freeman, Derek. 1983. *Margaret Mead and Samoa: The Making and Unmaking of an Anthropological Myth.* New York: Penguin.

Freeman, Derek. 1999. *The Fateful Hoaxing of Margaret Mead: A Historical Analysis of Her Samoan Research.* Boulder: Westview.

Freese, Jeremy and Brian Powell. 1999. "Sociobiology, Status, and Parental Investment in Sons and Daughters: Testing the Trivers-Willard Hypothesis." *American Journal of Sociology.* 106: 1704–43.

Friedman, Thomas. 2002. *Longitudes and Attitudes: Exploring the World Before and After September 11.* London: Penguin.

Frisch, Rose E. and Roger Revelle. 1970. "Height and Weight at Menarche and a Hypothesis of Critical Body Weights and Adolescent Events." *Science.* 169: 397–8.

Furchtgott-Roth, Diana and Christine Stolba. 1999. *Women's Figures: An Illustrated Guide to the Economic Progress of Women in America.* Washington: AEI Press.

Galef, Jr., Bennet G. and David J. White. 1998. "Mate-Choice Copying in Japanese Quail, *Coturnix coturnix japonica.*" *Animal Behaviour.* 55: 545–52.

Gallup, Jr., Gordon G., Rebecca L. Burch, Mary L. Zappieri, Rizwan A. Parvez, Malinda L. Stockwell, and Jennifer A. Davis. 2003. "The Human Penis as a Semen Displacement Device." *Evolution and Human Behavior.* 24: 277–89.

Gambetta, Diego. 2005. "Can We Make Sense of Suicide Missions?" Pp. 259–99 in *Making Sense of Suicide Missions*, edited by Diego Gambetta. Oxford: Oxford University Press.

Gangestad, Steven W. and David M. Buss. 1993. "Pathogen Prevalence and Human Mate Preferences." *Ethology and Sociobiology.* 14: 89–96.

Gangestad, Steven W. and Jeffry A. Simpson. 2000. "The Evolution of Human Mating: Trade-Offs and Strategic Pluralism." *Behavioral and Brain Sciences.* 23: 573–644.

Gangestad, Steven W. and Randy Thornhill. 1997. "The Evolutionary Psychology of Extrapair Sex: The Role of Fluctuating Asymmetry." *Evolution and Human Behavior.* 18: 69–88.

Gangestad, Steven W., Randy Thornhill, and Ronald A. Yeo. 1994. "Facial Attractiveness, Developmental Stability, and Fluctuating Asymmetry." *Ethology and Sociobiology.* 15: 73–85.

Gaulin, Steven J. C., Donald H. McBurney, and Stephanie L. Brakeman-Wartell. 1997. "Matrilateral Biases in the Investment of Aunts and Uncles: A Consequence and Measure of Paternity Uncertainty." *Human Nature.* 8: 139–51.

Gaulin, Steven J. C. and Carole J. Robbins. 1991. "Trivers-Willard Effect in Contemporary North American Society." *American Journal of Physical Anthropology.* 85: 61–9.

Gillis, John S. and Walter E. Avis. 1980. "The Male-Taller Norm in Mate Selection." *Personality and Social Psychology Bulletin.* 6: 396–401.

Glock, Charles, Benjamin Ringer, and Earl Babbie. 1967. *To Comfort and to Challenge.* Berkeley: University of California Press.

Gottschall, Jonathan, Johanna Martin, Hadley Quish, and Jon Rea. 2004. "Sex Differences in Mate Choice Criteria Are Reflected in Folktales from Around the World and in Historical European Literature." *Evolution and Human Behavior.* 25: 102–12.

Gould, Stephen J. and Richard C. Lewontin. 1979. The Spandrels of San Marcos and the Panglossian Paradigm: A Critique of the Adaptationist Programme. *Proceedings of the Royal Society of London, Series B.* 205: 581–98.

Grammer, Karl and Randy Thornhill. 1994. "Human (*Homo sapiens*) Facial Attractiveness and Sexual Selection: The Role of Symmetry and Averageness." *Journal of Comparative Psychology.* 108: 233–42.

Grant, J. W. A. and L. D. Green. 1996. "Mate Copying Versus Preference for Actively Courting Males by Female Japanese Medaka (*Oryzias latipes*)." *Behavioral Ecology.* 7: 165–7.

Greeley, Andrew. 1972. *The Denominational Society: A Sociological Approach to Religion in America.* Glenview: Scott Foresman.

Greenberg, David F. 1985. "Age, Crime, and Social Explanation." *American Journal of Sociology.* 91: 1–21.

Guthrie, Stewart Elliott. 1993. *Faces in the Clouds: A New Theory of Religion.* New York: Oxford University Press.

Hall, Daniel E. 2006. "Religious Attendance: More Cost-Effective Than Lipitor?" *Journal of the American Board of Family Medicine.* 19: 103–9.

Hamer, Dean H. 2004. *The God Gene: How Faith Is Hardwired into Our Genes.* New York: Doubleday.

Hamer, Dean H. and Peter Copeland. 1994. *The Science of Desire: The Search for the Gay Gene and the Biology of Behavior.* New York: Simon and Schuster.

Hamer, Dean H., Stella Hu, Victoria L. Magnuson, Nan Hu, and Angela M. L. Pattatucci. 1993. "A Linkage Between DNA Markers on the X Chromosome and Male Sexual Orientation." *Science.* 261: 321–7.

Hargens, Lowell L., James C. McCann, and Barbara F. Reskin. 1978. "Productivity and Reproductivity: Fertility and Professional Achievement among Research Scientists." *Social Forces.* 57: 154–63.

Harris, Christine R. 2003. "A Review of Sex Differences in Sexual Jealousy, Including Self-Report Data, Psychophysiological Responses, Interpersonal Violence, and Morbid Jealousy." *Personality and Social Psychology Review.* 7: 102–28.

Harris, Judith Rich. 1995. "Where Is the Child's Environment?: A Group Socialization Theory of Development." *Psychological Review.* 102: 458–89.

Harris, Judith Rich. 1998. *The Nurture Assumption: Why Children Turn Out the Way They Do.* New York: Free Press.

Harris, Marvin. 1974. *Cows, Pigs, Wars and Witches: The Riddles of Culture.* New York: Random House.

Haselton, Martie G. 2003. "The Sexual Overperception Bias: Evidence of a Systematic Bias in Men from a Survey of Naturally Occurring Events." *Journal of Research in Personality.* 37: 34–47.

Haselton, Martie G. and David M. Buss. 2000. "Error Management Theory: A New Perspective on Biases in Cross-Sex Mind Reading." *Journal of Personality and Social Psychology.* 78: 81–91.

Haselton, Martie G. and Daniel Nettle. 2006. "The Paranoid Optimist: An Integrative Evolutionary Model of Cognitive Biases." *Personality and Social Psychology Review.* 10: 47–66.

Hechter, Michael. 2000. *Containing Nationalism.* Oxford: Oxford University Press.

Hejl, Peter M., Manfred Kammer, and Matthias Uhl. Forthcoming. "The Really Interesting Stories Are the Old Ones: Evolved Interests in Economically Successful Films from Hollywood and Bollywood." In *You Can't Turn It Off: Media, Mind, and Evolution*, edited by Jerome H. Barkow and Peter M. Hejl. New York: Oxford University Press.

Helm, P., K. R. Münster, and L. Schmidt. 1995. "Recalled Menarche in Relation to Infertility and Adult Weight and Height." *Acta Obstetricia Et Gynecologica Scandinavica.* 74: 718–22.

Hemley, Robin. 2003. *Invented Eden: The Elusive, Disputed History of the Tasaday.* New York: Farrar, Straus and Giroux.

Henderson, Joshua J. A. and Jeremy M. Anglin. 2003. "Facial Attractiveness Predicts Longevity." *Evolution and Human Behavior.* 24: 351–6.

Herman-Giddens, M. E., E. J. Slora, R. C. Wasserman, C. J. Bourdony, M. V. Bhapkar, G. G. Koch, and C. M. Hasemeier. 1997. "Secondary Sexual Characteristics and Menses in Young Girls Seen in Office Practice: A Study from the Pediatric Research in Office Settings Network." *Pediatrics.* 99: 505–12.

Hess, Eckhard H. 1975. *The Tell-Tale Eye: How Your Eyes Reveal Hidden Thoughts and Emotions.* New York: Van Nostrand Reinhold.

Hess, Eckhard H. and James M. Polt. 1960. "Pupil Size Related to Interest Value of Visual Stimuli." *Science.* 132: 349–50.

Hirschi, Travis. 1969. *Causes of Delinquency.* Berkeley: University of California Press.

Hirschi, Travis and Michael Gottfredson. 1985. "Age and Crime, Logic and Scholarship: Comment on Greenberg." *American Journal of Sociology.* 91: 22–7.

Höglund, Jacob, Rauno V. Alatalo. Robert M. Gibson, and Arne Lundberg. 1995. "Mate-Choice Copying in Black Grouse." *Animal Behaviour.* 49: 1627–33.

Hönekopp, Johannes, Tobias Bartholomé, and Gregor Jansen. 2004. "Facial Attractiveness, Symmetry, and Physical Fitness in Young Women." *Human Nature.* 15: 147–67.

Hume, David. 1739. *A Treatise of Human Nature: Being an Attempt to Introduce the Experimental Method of Reasoning into Moral Subjects and Dialogues Concerning Natural Religion.* London: John Noon.

Iannaccone, Laurence R. 1990. "Religious Practice: A Human Capital Approach." *Journal for the Scientific Study of Religion.* 29: 297–314.

Iannaccone, Laurence R. 1994. "Why Strict Churches Are Strong." *American Journal of Sociology.* 99: 1180–1211.

International Criminal Police Organization. Various years. *International Criminal Statistics.* Lyon: Interpol.

James, William. 1985 [1902]. *Varieties of Religious Experience.* Cambridge: Harvard University Press.

Jaruratanasirikul, S., L. Mo-suwan, and L. Lebel. 1997. "Growth Pattern and Age at Menarche of Obese Girls in a Transitional Society." *Journal of Pediatric Endocrinology and Metabolism.* 10: 487–90.

Jasienska, Grazyna, Anna Ziomkiewicz, Peter T. Ellison, Susan F. Lipton, and Inger Thune. 2004. "Large Breasts and Narrow Waists Indicate High Reproductive Potential in Women." *Proceedings of the Royal Society of London, Series B.* 271: 1213–17.

Jones, Doug. 1996. *Physical Attractiveness and the Theory of Sexual Selection.* Ann Arbor: University of Michigan Museum of Anthropology.

Jones, Doug and Kim Hill. 1993. "Criteria of Physical Attractiveness in Five Populations." *Human Nature.* 4: 271–96.

Kalick, S. Michael, Leslie A. Zebrowitz, Judith H. Langlois, and Robert M. Johnson. 1998. "Does Human Facial Attractiveness Honestly Advertise Health?: Longitudinal Data on an Evolutionary Question." *Psychological Science.* 9: 8–13.

Kalyvas, Stathis N. 2005. "Warfare in Civil Wars." Pp. 88–108 in *Rethinking the Nature of War,* edited by Isabelle Duyvesteyn and Jan Angstrom. Abington: Frank Cass.

Kanazawa, Satoshi. 2000. "Scientific Discoveries as Cultural Displays: A Further Test of Miller's Courtship Model." *Evolution and Human Behavior.* 21: 317–21.

Kanazawa, Satoshi. 2001a. "De Gustibus Est Disputandum." *Social Forces.* 79: 1131–63.

Kanazawa, Satoshi. 2001b. "Where Do Social Structures Come From?" *Advances in Group Processes.* 18: 161–83.

Kanazawa, Satoshi. 2001c. "Why Father Absence Might Precipitate Early Menarche: The Role of Polygyny." *Evolution and Human Behavior.* 22: 329–34.

Kanazawa, Satoshi. 2001d. "Why We Love Our Children." *American Journal of Sociology.* 106: 1761–76.

Kanazawa, Satoshi. 2002. "Bowling with Our Imaginary Friends." *Evolution and Human Behavior.* 23: 167–71.

Kanazawa, Satoshi. 2003a. "Can Evolutionary Psychology Explain Reproductive Behavior in the Contemporary United States?" *Sociological Quarterly.* 44: 291–302.

Kanazawa, Satoshi. 2003b. "A General Evolutionary Psychological Theory of Male Criminality and Related Male-Typical Behavior." Pp. 37–60 in *Biosocial Criminology: Challenging Environmentalism's Supremacy,* edited by Anthony Walsh and Lee Ellis. New York: Nova Science Publishers.

Kanazawa, Satoshi. 2003c. "Why Productivity Fades with Age: The Crime-Genius Connection." *Journal of Research in Personality.* 37: 257–72.

Kanazawa, Satoshi. 2004a. "General Intelligence as a Domain-Specific Adaptation." *Psychological Review.* 111: 512–23.

Kanazawa, Satoshi. 2004b. "The Savanna Principle." *Managerial and Decision Economics.* 25: 41–54.

Kanazawa, Satoshi. 2004c. "Social Sciences Are Branches of Biology." *Socio-Economic Review.* 2: 371–90.

Kanazawa, Satoshi. 2005a. "Big and Tall Parents Have More Sons: Further Generalizations of the Trivers-Willard Hypothesis." *Journal of Theoretical Biology.* 235: 583–90.

Kanazawa, Satoshi. 2005b. "Is 'Discrimination' Necessary to Explain the Sex Gap in Earnings?" *Journal of Economic Psychology.* 26: 269–87.

Kanazawa, Satoshi. 2006a. "'First, Kill All the Economists. . . .': The Insufficiency of Microeconomics and the Need for Evolutionary Psychology in the Study of Management." *Managerial and Decision Economics.* 27: 95–101.

Kanazawa, Satoshi. 2006b. "Violent Men Have More Sons: Further Evidence for the Generalized Trivers-Willard Hypothesis (gTWH)." *Journal of Theoretical Biology.* 239: 450–59.

Kanazawa, Satoshi. 2006c. "Where Do Cultures Come From?" *Cross-Cultural Research.* 40: 152–76.

Kanazawa, Satoshi. 2007. "Beautiful Parents Have More Daughters: A Further Implication of the Generalized Trivers-Willard Hypothesis (gTWH)." *Journal of Theoretical Biology.* 244: 133–40.

Kanazawa, Satoshi and Rebecca L. Frerichs. 2001. "Why Single Men Might Abhor Foreign Cultures." *Social Biology.* 48: 320–7.

Kanazawa, Satoshi and Deanna L. Novak. 2005. "Human Sexual Dimorphism in Size May Be Triggered by Environmental Cues." *Journal of Biosocial Science.* 37: 657–65.

Kanazawa, Satoshi and Mary C. Still. 1999. "Why Monogamy?" *Social Forces.* 78: 25–50.

Kanazawa, Satoshi and Mary C. Still. 2000a. "Parental Investment as a Game of Chicken." *Politics and the Life Sciences.* 19: 17–26.

Kanazawa, Satoshi and Mary C. Still. 2000b. "Teaching May Be Hazardous to Your Marriage." *Evolution and Human Behavior.* 21: 185–90.

Kanazawa, Satoshi and Mary C. Still. 2000c. "Why Men Commit Crimes (and Why They Desist)." *Sociological Theory.* 18: 434–47.

Kanazawa, Satoshi and Mary C. Still. 2001. "The Emergence of Marriage

Norms: An Evolutionary Psychological Perspective." Pp. 274–304 in *Social Norms*, edited by Michael Hechter and Karl-Dieter Opp. New York: Russell Sage Foundation.

Kanazawa, Satoshi and Griet Vandermassen. 2005. "Engineers Have More Sons, Nurses Have More Daughters: An Evolutionary Psychological Extension of Baron-Cohen's Extreme Male Brain Theory of Autism and Its Empirical Implications." *Journal of Theoretical Biology*. 233: 589–99.

Kaprio, Jaakko, Arja Rimpela, Torsten Winter, Richard J. Viken, Matti Rimpela, and Richard J. Rose. 1995. "Common Genetic Influences on BMI and Age at Menarche." *Human Biology*. 67: 739–53.

Katzev, Aphra R., Rebecca L. Warner, and Alan C. Acock. 1994. "Girls or Boys? Relationship of Child Gender to Marital Stability." *Journal of Marriage and the Family*. 56: 89–100.

Keller, Matthew C., Randolph M. Nesse, and Sandra Hofferth. 2001. "The Trivers-Willard Hypothesis of Parental Investment: No Effect in the Contemporary United States." *Evolution and Human Behavior*. 22: 343–60.

Kenrick, Douglas T. and Richard C. Keefe. 1992. "Age Preferences in Mates Reflect Sex Differences in Reproductive Strategies." *Behavioral and Brain Sciences*. 15: 75–133.

Ketelaar, Timothy and Bruce J. Ellis. 2000. "Are Evolutionary Explanations Unfalsifiable? Evolutionary Psychology and the Lakatosian Philosophy of Science." *Psychological Inquiry*. 11: 1–21.

Kirkpatrick, Lee A. 2005. *Attachment, Evolution, and the Psychology of Religion*. New York: Guilford.

Kirkpatrick, Mark. 1987. "Sexual Selection by Female Choice in Polygynous Animals." *Annual Review of Ecology and Systematics*. 18: 43–70.

Koenig, Laura B., Matt McGue, Robert F. Krueger, and Thomas J. Bouchard, Jr. 2005. "Genetic and Environmental Influences on Religiousness: Findings for Retrospective and Current Religiousness Ratings." *Journal of Personality*. 73: 471–88.

Kohler, Hans-Peter, Joseph L. Rodgers, and Kaare Christensen. 1999. "Is Fertility Behavior in Our Genes? Findings from a Danish Twin Study." *Population and Development Review*. 25: 253–88.

Krueger, Alan B. and Jitka Maleckova. 2003. "Education, Poverty and Terrorism: Is There a Causal Connection?" *Journal of Economic Perspectives*. 17: 119–44.

Kurzban, Robert and Martie G. Haselton. 2006. "Making Hay Out of Straw? Real and Imagined Controversies in Evolutionary Psychology." Pp. 149–61 in

*Missing the Revolution: Darwinism for Social Scientists.* Oxford: Oxford University Press.

Kurzban, Robert, John Tooby, and Leda Cosmides. 2001. "Can Race Be Erased? Coalitional Computation and Social Categorization." *Proceedings of the National Academy of Sciences.* 98: 15387–92.

Langlois, Judith H., Lisa Kalakanis, Adam J. Rubenstein, Andrea Larson, Monica Hallam, and Monica Smoot. 2000. "Maxims or Myths of Beauty?: A Meta-Analytic and Theoretical Review." *Psychological Bulletin.* 126: 390–423.

Langlois, Judith H. and Lori A. Roggman. 1990. "Attractive Faces Are Only Average." *Psychological Science.* 1: 115–21.

Langlois, Judith H., Lori A. Roggman, Rita J. Casey, Jean M. Ritter, Loretta A. Rieser-Danner, and Vivian Y. Jenkins. 1987. "Infant Preferences for Attractive Faces: Rudiments of a Stereotype?" *Developmental Psychology.* 23: 363–9.

Langlois, Judith H., Lori A. Roggman, and Lisa Musselman, 1994. "What Is Average and What Is Not Average about Attractive Faces?" *Psychological Science.* 5: 214–20.

Langlois, Judith H., Lori A. Roggman, and Loretta A. Rieser-Danner. 1990. "Infants' Differential Social Responses to Attractive and Unattractive Faces." *Developmental Psychology.* 26: 153–9.

Laub, John H., Daniel S. Nagin, and Robert J. Sampson. 1998. "Trajectories of Change in Criminal Offending: Good Marriages and the Desistance Process." *American Sociological Review.* 63: 225–38.

Laumann, Edward O., John H. Gagnon, Robert T. Michael, and Stuart Michaels. 1994. *The Social Organization of Sexuality: Sexual Practices in the United States.* University of Chicago Press.

Lemonick, Michael D. 2000. "Teens Before Their Time." *Time.* 156 (18): 66–74.

Lenski, Gerhard E. 1966. *Power and Privilege: A Theory of Social Stratification.* Chapel Hill: University of North Carolina Press.

Leutenegger, Walter and James T. Kelly. 1977. "Relationship of Sexual Dimorphism in Canine Size and Body Size to Social, Behavioral, and Ecological Correlates in Anthropoid Primates." *Primates.* 18: 117–36.

Liedtke, Michael. 2000. "Smiles More Discerning at Safeway." *Contra Costa Times.* January 18. Business and Financial News Section.

Liss, Lora. 1987. "Families and the Law." Pp. 767–93 in *Handbook of Marriage and the Family,* edited by Marvin B. Sussman and Suzanne K. Steinmetz. New York: Plenum.

Little, Anthony C., Ian S. Penton-Voak, D. Michael Burt, and David I. Perrett.

2002. "Evolution and Individual Differences in the Perception of Attractiveness: How Cyclic Hormonal Changes and Self-Perceived Attractiveness Influence Female Preferences for Male Faces." Pp. 59–90 in *Facial Attractiveness: Evolutionary, Cognitive, and Social Perspectives*, edited by Gillian Rhodes and Leslie A. Zebrowitz. Westport: Ablex.

Low, Bobbi S. 1979. "Sexual Selection and Human Ornamentation." Pp. 462–87 in *Evolutionary Biology and Human Social Behavior: An Anthropological Perspective*, edited by Napoleon A. Chagnon and William Irons. North Scituate: Duxbury.

Luckmann, Thomas. 1967. *The Invisible Religion*. New York: Macmillan.

Lycett, J. E. and R. I. M. Dunbar. 2000. "Mobile Phones as Lekking Devices Among Human Males." *Human Nature*. 11: 93–104.

Machalek, Richard and Michael W. Martin. 2004. "Sociology and the Second Darwinian Revolution: A Metatheoretical Analysis." *Sociological Theory*. 22: 455–76.

Maret, Stephen M. and Craig A. Harling. 1985. "Cross-Cultural Perceptions of Physical Attractiveness: Ratings of Photographs of Whites by Cruzans and Americans." *Perceptual and Motor Skills*. 60: 163–6.

Marini, Margaret Mooney. 1989. "Sex Differences in Earnings in the United States." *Annual Review of Sociology*. 15: 343–80.

Marlowe, Frank. 1998. "The Nubility Hypothesis: The Human Breast as an Honest Signal of Residual Reproductive Value." *Human Nature*. 9: 263–71.

Marsden, Peter V. 1987. "Core Discussion Networks of Americans." *American Sociological Review*. 52: 122–31.

Martin, David. 1967. *A Sociology of English Religion*. London: SCM Press.

Maynard Smith, John. 1997. "Commentary." Pp. 522–6 in *Feminism and Evolutionary Biology: Boundaries, Intersections, and Frontiers*, edited by Patricia Adair Gowaty. New York: Chapman and Hall.

McCullough, Michael E., William T. Hoyt, David B. Larson, Harold G. Koenig, and Carl Thoresen. 2000. "Religious Involvement and Mortality: A Meta-Analytic Review." *Health Psychology*. 19: 211–22.

McLain, D. Kelly, Deanna Setters, Michael P. Moulton, and Ann E. Pratt. 2000. "Ascription of Resemblance of Newborns by Parents and Nonrelatives." *Evolution and Human Behavior*. 21: 11–23.

McNamara, Patrick. 2001. "Religion and the Frontal Lobes." Pp. 237–56 in *Religion in Mind: Cognitive Perspectives on Religious Belief, Ritual, and Experience*, edited by Jensine Andresen. New York: Cambridge University Press.

McWhirter, Norris and Ross McWhirter. 1975. *The Guinness Book of World Records 1976*. New York: Sterling.

Mealey, Linda. 1992. "Alternative Adaptive Models of Rape." *Behavioral and Brain Sciences.* 15: 397–8.

Mealey, Linda. 2000. *Sex Differences: Development and Evolutionary Strategies.* San Diego: Academic Press.

Mealey, L., R. Bridgstock, and G. C. Townsend. 1999. "Symmetry and Perceived Facial Attractiveness: A Monozygotic Co-Twin Comparison." *Journal of Personality and Social Psychology.* 76: 151–8.

Mesko, Norbert and Tamas Bereczkei. 2004. "Hairstyle as an Adaptive Means of Displaying Phenotypic Quality." *Human Nature.* 15: 251–70.

Miller, Alan S. and John P. Hoffmann. 1995. "Risk and Religion: An Explanation of Gender Differences in Religiosity." *Journal for the Scientific Study of Religion.* 34: 63–75.

Miller, Alan S. and Satoshi Kanazawa. 2000. *Order by Accident: The Origins and Consequences of Conformity in Contemporary Japan.* Boulder: Westview.

Miller, Alan S. and Rodney Stark. 2002. "Gender and Religiousness: Can Socialization Explanations Be Saved?" *American Journal of Sociology.* 107: 1399–423.

Miller, Geoffrey F. 1998. "How Mate Choice Shaped Human Nature: A Review of Sexual Selection and Human Evolution." Pp. 87–129 in *Handbook of Evolutionary Psychology: Ideas, Issues, and Applications*, edited by C. Crawford and D. L. Krebs. Lawrence Erlbaum.

Miller, Geoffrey F. 1999. "Sexual Selection for Cultural Displays." Pp. 71–91 in *The Evolution of Culture: An Interdisciplinary View*, edited by Robin Dunbar, Chris Knight, and Camilla Power. New Brunswick: Rutgers University Press.

Miller, Geoffrey F. 2000. *The Mating Mind: How Sexual Choice Shaped the Evolution of Human Nature.* New York: Doubleday.

Miller, William R. and Carl E. Thoresen. 2003. "Spirituality, Religion, and Health: An Emerging Research Field." *American Psychologist.* 58: 24–35.

Moir, Anne and David Jessel. 1989. *Brain Sex: The Real Difference Between Men and Women.* New York: Delta.

Mol, Hans. 1985. *The Faith of Australians.* Sydney: George, Allen and Unwin.

Møller, Anders Pape. 1990. "Fluctuating Asymmetry in Male Sexual Ornaments May Reliably Reveal Male Quality." *Animal Behaviour.* 40: 1185–7.

Møller, Anders Pape. 1992. "Parasites Differentially Increase the Degree of Fluctuating Asymmetry in Secondary Sexual Characters." *Journal of Evolutionary Biology.* 5: 691–700.

Moore, George Edward. 1903. *Principia Ethica.* Cambridge: Cambridge University Press.

Moore, John H. 1990. "The Reproductive Success of Cheyenne War Chiefs: A Contrary Case to Chagnon's Yanomamö." *Current Anthropology.* 31: 322–30.

Morgan, S. Philip, Diane N. Lye, and Gretchen A. Condran. 1988. "Sons, Daughters, and the Risk of Marital Disruption." *American Journal of Sociology.* 94: 110–29.

Morse, Stanley J. and Joan Gruzen. 1976. "The Eye of the Beholder: A Neglected Variable in the Study of Physical Attractiveness?" *Journal of Personality.* 44: 209–25.

Muehlenhard, Charlene L. and Lisa C. Hollabaugh. 1988. "Do Women Sometimes Say No When They Mean Yes? The Prevalence and Correlates of Women's Token Resistance to Sex." *Journal of Personality and Social Psychology.* 54: 872–9.

Muehlenhard, Charlene L. and Marcia L. McCoy. 1991. "Double Standard/Double Bind: The Sexual Double Standard and Women's Communication about Sex." *Psychology of Women Quarterly.* 15: 447–61.

Mueller, Charles W., Sarosh Kuruvilla, and Roderick D. Iverson. 1994. "Swedish Professionals and Gender Inequalities." *Social Forces.* 73: 555–73.

Mueller, Ulrich. 1993. "Social Status and Sex." *Nature.* 363: 490.

Munch, Allison, J. Miller McPherson, and Lynn Smith-Lovin. 1997. "Gender, Children, and Social Contact: The Effects of Childrearing for Men and Women." *American Sociological Review.* 62: 509–20.

Nance, John. 1975. *The Gentle Tasaday: A Stone Age People in the Philippine Rain Forest.* New York: Harcourt Brace Jovanovich.

Nettle, Daniel. 2002. "Women's Height, Reproductive Success and the Evolution of Sexual Dimorphism in Modern Humans." *Proceedings of the Royal Society of London, Series B–Biological Sciences.* 269: 1919–23.

Newberg, Andrew, Eugene D'Aquili, and Vince Rause. 2002. *Why God Won't Go Away: Brain Science and the Biology of Belief.* New York: Ballantine.

O'Hanlon, Michael E. and Jason H. Campbell. 2007. Iraq Index: Tracking Variables of Reconstruction & Security in Post-Saddam Iraq. Washington, DC: Brookings Institution. [www.brookings.edu/fp/saban/iraq/indexarchive.htm]

Okasha, M., P. McCarron, J. McEwen, and G. Davey Smith. 2001. "Age at Menarche: Secular Trends and Association with Adult Anthropometric Measures." *Annals of Human Biology.* 28: 68–78.

Olson, Mancur. 1965. *The Logic of Collective Action: Public Goods and the Theory of Groups.* Cambridge: Harvard University Press.

Orians, Gordon H. 1969. "On the Evolution of Mating Systems in Birds and Mammals." *American Naturalist.* 103: 589–603.

Paludi, Michele. (Editor.) 1996. *Sexual Harassment on College Campuses: Abusing the Ivory Power.* Albany: SUNY Press.

Parsons, P. A. 1990. "Fluctuating Asymmetry: An Epigenetic Measure of Stress." *Biological Review.* 65: 131–45.

Parsons, P. A. 1992. "Fluctuating Asymmetry: A Biological Monitor of Environmental and Genomic Stress." *Heredity.* 68: 361–4.

Pate, Kelly. 2001. "Seller Beware! Secret Shoppers Check Service; Some Call It Spying." *Denver Post.* July 1. Business Section, K-01.

Perrett, David I., Michael Burt, Ian S. Penton-Voak, Kieran J. Lee, Duncan A. Rowland, and Rachel Edwards. 1999. "Symmetry and Human Facial Attractiveness." *Evolution and Human Behavior.* 20: 295–307.

Pérusse, Daniel. 1993. "Cultural and Reproductive Success in Industrial Societies: Testing the Relationship at the Proximate and Ultimate Levels." *Behavioral and Brain Sciences.* 16: 267–322.

Pérusse, Daniel. 1994. "Mate Choice in Modern Societies: Testing Evolutionary Hypotheses with Behavioral Data." *Human Nature.* 5: 255–78.

Peters, Jay, Todd K. Shackelford, and David M. Buss. 2002. "Understanding Domestic Violence Against Women: Using Evolutionary Psychology to Extend the Feminist Functional Analysis." *Violence and Victims.* 17: 255–64.

Pietrzak, Robert H., James D. Laird, David A. Stevens, and Nicholas S. Thompson. 2002. "Sex Differences in Human Jealousy: A Coordinated Study of Forced-Choice, Continuous Rating-Scale, and Physiological Responses on the Same Subjects." *Evolution and Human Behavior.* 23: 83–94.

Pinker, Steven. 1994. *The Language Instinct.* New York: HarperCollins.

Pinker, Steven. 2002. *The Blank Slate: The Modern Denial of Human Nature.* London: Penguin.

Quinlan, Robert J. 2003. "Father Absence, Parental Care, and Female Reproductive Development." *Evolution and Human Behavior.* 24: 376–90.

Ramanchandran, V. S. 1997. "Why Do Gentlemen Prefer Blondes?" *Medical Hypotheses.* 48: 19–20.

Rand, Colleen S. and John M. Kuldau. 1990. "The Epidemiology of Obesity and Self-Defined Weight Problem in the General Population: Gender, Race, Age, and Social Class." *International Journal of Eating Disorders.* 9: 329–43.

Ream, Sarah L. 2000. "When Service with a Smile Invites More Than Satisfied

Customers: Third-Party Sexual Harassment and the Implications of Charges Against Safeway." *Hastings Women's Law Journal.* 11: 107–22.

Regalski, Jeanne M. and Steven J. C. Gaulin. 1993. "Whom Are Mexican Infants Said to Resemble?: Monitoring and Fostering Paternal Confidence in the Yucatan." *Ethology and Sociobiology.* 14: 97–113.

Rhodes, Gillian, Leigh W. Simmons, and Marianne Peters. 2005. "Attractiveness and Sexual Behavior: Does Attractiveness Enhance Mating Success?" *Evolution and Human Behavior.* 26: 186–201.

Rich, Melissa K. and Thomas F. Cash. 1993. "The American Image of Beauty: Media Representations of Hair Color for Four Decades." *Sex Roles.* 29: 113–24.

Ridley, Matt. 1993. *The Red Queen: Sex and the Evolution of Human Nature.* New York: Penguin.

Ridley, Matt. 1996. *The Origins of Virtue: Human Instincts and the Evolution of Cooperation.* New York: Viking.

Ridley, Matt. 1999. *Genome: The Autobiography of a Species in 23 Chapters.* New York: Perennial.

Rodgers, Joseph Lee, Kimberly Hughes, Hans-Peter Kohler, Kaare Christensen, Debby Doughty, David C. Rowe, and Warren B. Miller. 2001. "Genetic Influence Helps Explain Variation in Human Fertility: Evidence from Recent Behavioral and Molecular Genetic Studies." *Current Directions in Psychological Science.* 10: 184–8.

Rodgers, Joseph Lee, Hans-Peter Kohler, Kirsten Ohm Kyvik, and Kaare Christensen. 2001. "Behavior Genetic Modeling of Human Fertility: Findings from a Contemporary Danish Study." *Demography.* 38: 29–42.

Rosenfeld, Rachel A. and Arne L. Kalleberg. 1990. "A Cross-National Comparison of Gender Gap in Income." *American Journal of Sociology.* 96: 69–106.

Rowe, David C. 1994. *The Limits of Family Influence: Genes, Experience, and Behavior.* New York: Guilford.

Rowe, David C. 2002. "On Genetic Variation in Menarche and Age at First Sexual Intercourse: A Critique of the Belsky-Draper Hypothesis." *Evolution and Human Behavior.* 23: 365–72.

Rubenstein, Adam J., Judith H. Langlois, and Lori A. Roggman. 2002. "What Makes a Face Attractive and Why: The Role of Averageness in Defining Facial Beauty." Pp. 1–33 in *Facial Attractiveness: Evolutionary, Cognitive, and Social Perspectives,* edited by Gillian Rhodes and Leslie A. Zebrowitz. Westport: Ablex.

Saad, Gad and Albert Peng. 2006. "Applying Darwinian Principles in Designing

Effective Intervention Strategies: The Case of Sun Tanning." *Psychology and Marketing.* 23: 617–38.

Salmon, Catherine and Donald Symons. 2001. *Warrior Lovers: Erotic Fiction, Evolution and Female Sexuality.* London: Weidenfeld and Nicolson.

Salmon, Catherine and Donald Symons. 2004. "Slash Fiction and Human Mating Psychology." *Journal of Sex Research.* 41: 94–100.

Sampson, Robert J. and John H. Laub. 1993. *Crime in the Making: Pathways and Turning Points Through Life.* Cambridge: Harvard University Press.

Samuels, Curtis A. and Richard Ewy. 1985. "Aesthetic Perception of Faces During Infancy." *British Journal of Developmental Psychology.* 3: 221–8.

Scarr, Sandra. 1995. "Psychology Will Be Truly Evolutionary When Behavior Genetics Is Included." *Psychological Inquiry.* 6: 68–71.

Schmitt, David P. 2003. "Universal Sex Differences in the Desire for Sexual Variety: Tests from 52 Nations, 6 Continents, and 13 Islands." *Journal of Personality and Social Psychology.* 83: 85–104.

Schmitt, David P. 2004. "Pattern and Universals of Mate Poaching Across 53 Nations: The Effects of Sex, Culture, and Personality on Romantically Attracting Another Person's Partner." *Journal of Personality and Social Psychology.* 86: 560–84.

Searcy, William A. and Ken Yasukawa. 1989. "Alternative Models of Territorial Polygyny in Birds." *American Naturalist.* 134: 323–43.

Shackelford, Todd K. and Randy J. Larsen. 1999. "Facial Attractiveness and Physical Health." *Evolution and Human Behavior.* 20: 71–6.

Shaw, Bernard. 1957. *Man and Superman.* New York: Penguin.

Sheppard, James A. and Alan J. Strathman. 1989. "Attractiveness and Height: The Role of Stature in Dating Preference, Frequency of Dating, and Perceptions of Attractiveness." *Personality and Social Psychology Bulletin.* 15: 617–27.

Sherkat, Darren E. 2002. "Sexuality and Religious Commitment: An Empirical Assessment." *Journal for the Scientific Study of Religion.* 41: 313–23.

Shields, William M. and Lea M. Shields. 1983. "Forcible Rape: An Evolutionary Analysis." *Ethology and Sociobiology.* 4: 115–36.

Silventoinen, Karri, Jaakko Kaprio, Eero Lahelma, Richard J. Viken, and Richard J. Rose. 2001. "Sex Differences in Genetic and Environmental Factors Contributing to Body-Height." *Twin Research.* 4: 25–9.

Singh, Devendra. 1993. "Adaptive Significance of Waist-to-Hip Ratio and Female Physical Attractiveness." *Journal of Personality and Social Psychology.* 65: 293–307.

Singh, D. and S. Luis. 1995. "Ethnic and Gender Consensus for the Effect of

Waist-to-Hip Ratio on Judgments of Women's Attractiveness." *Human Nature.* 6: 51–65.

Singh, D. and R. K. Young. 1995. "Body Weight, Waist-to-Hip Ratio, Breasts, and Hips: Role in Judgments of Female Attractiveness and Desirability for Relationships." *Ethology and Sociobiology.* 16: 483–507.

Slater, Alan, Charlotte Von der Schulenburg, Elizabeth Brown, Marion Badenoch, George Butterworth, Sonia Parsons, and Curtis Samuels. 1998. "Newborn Infants Prefer Attractive Faces." *Infant Behavior and Development.* 21: 345–54.

Small, Meredith F. 1993. *Female Choices: Sexual Behavior of Female Primates.* Cornell University Press.

Smith, Robert L. 1984. "Human Sperm Competition." Pp. 601–59 in *Sperm Competition and the Evolution of Mating Systems,* edited by Robert L. Smith. New York: Academic Press.

Smith-Lovin, Lynn and J. Miller McPherson. 1993. "You Are Who You Know: A Network Approach to Gender." Pp. 223–51 in *Theory on Gender/Feminism on Theory,* edited by Paula England. New York: Aldine.

Sørensen, Annemette and Heike Trappe. 1995. "The Persistence of Gender Inequality in Earnings in the German Democratic Republic." *American Sociological Review.* 60: 398–406.

Sozou, Peter D. and Robert M. Seymour. 2005. "Costly but Worthless Gifts Facilitate Courtship." *Proceedings for the Royal Society of London, Series B.* 272: 1877–84.

Stark, Rodney. 1992. *Doing Sociology.* Belmont: Wadsworth.

Stark, Rodney. 2002. "Physiology and Faith: Addressing the 'Universal' Gender Difference in Religious Commitment." *Journal for the Scientific Study of Religion.* 41: 495–507.

Steffensmeier, Darrell J., Emilie Andersen Allan, Miles D. Harer, and Cathy Streifel. 1989. "Age and the Distribution of Crime." *American Journal of Sociology.* 94: 803–31.

Steggarda, M. 1993. "Religion and the Social Position of Men and Women." *Social Compass.* 40: 65–73.

Sulloway, Frank J. 1996. *Born to Rebel: Birth Order, Family Dynamics, and Creative Lives.* New York: Pantheon.

Sulloway, Frank J. 2000. "*Born to Rebel* and Its Critics." *Politics and the Life Sciences.* 19: 181–202.

Suziedalis, Antanas and Raymond H. Potvin. 1981. "Sex Differences in Factors Affecting Religiousness Among Catholic Adolescents." *Journal for Scientific Study of Religion.* 20: 38–50.

Symons, Donald. 1979. *The Evolution of Human Sexuality.* Oxford: Oxford University Press.

Symons, Donald. 1990. "Adaptiveness and Adaptation." *Ethology and Sociobiology.* 11: 427–44.

Symons, Donald. 1995. "Beauty Is in the Adaptations of the Beholder: The Evolutionary Psychology of Human Female Sexual Attractiveness." Pp. 80–118 in *Sexual Nature, Sexual Culture,* edited by Paul R. Abrahamson and Steven D. Pinkerton. Chicago: University of Chicago Press.

Takahashi, Chisato, Toshio Yamagishi, Shigehito Tanida, Toko Kiyonari, and Satoshi Kanazawa. 2006. "Attractiveness and Cooperation in Social Exchange." *Evolutionary Psychology.* 4: 315–29.

Tangri, Sandra S., Martha R. Burt, and Leanor B. Johnson. 1982. "Sexual Harassment at Work: Three Explanatory Models." *Journal of Social Issues.* 38: 33–54.

Thakerar, Jitendra N. and Saburo Iwawaki. 1979. "Cross-Cultural Comparisons in Interpersonal Attraction of Females Toward Males." *Journal of Social Psychology.* 108: 121–2.

Thiessen, Del and Yoko Umezawa. 1998. "The Sociobiology of Everyday Life: A New Look at a Very Old Novel." *Human Nature.* 9: 293–320.

Thornhill, Randy. 1976. "Sexual Selection and Nuptial Feeding Behavior in *Bittacus apicalis* (Insecta: Mecoptera)." *American Naturalist.* 119: 529–48.

Thornhill, Randy, and Steven W. Gangestad. 1993. "Human Facial Beauty: Averageness, Symmetry, and Parasite Resistance." *Human Nature.* 4: 237–69.

Thornhill, Randy and Anders Pape Møller. 1997. "Developmental Stability, Disease and Medicine." *Biological Review.* 72: 497–548.

Thornhill, Randy and Craig T. Palmer. 2000. *A Natural History of Rape: Biological Bases of Sexual Coercion.* Cambridge: MIT Press.

Thornhill, Randy and Nancy Wilmsen Thornhill. 1983. "Human Rape: An Evolutionary Analysis." *Ethology and Sociobiology.* 4: 137–73.

Tooby, John and Leda Cosmides. 1990. "The Past Explains the Present: Emotional Adaptations and the Structure of Ancestral Environments." *Ethology and Sociobiology.* 10: 29–49.

Tooby, John and Leda Cosmides. 1992. "The Psychological Foundations of Culture." Pp. 19–136 in *The Adapted Mind: Evolutionary Psychology and the Generation of Culture,* edited by Jerome H. Barkow, Leda Cosmides, and John Tooby. New York: Oxford University Press.

Townsend, John M. and Gary D. Levy. 1990. "Effects of Potential Partners' Costume and Physical Attractiveness on Sexuality and Partner Selection." *Journal of Psychology.* 124: 371–89.

Trivers, Robert L. 1972. "Parental Investment and Sexual Selection." Pp. 136–79 in *Sexual Selection and the Descent of Man 1871–1971*, edited by Bernard Campbell. Chicago: Aldine.

Trivers, Robert. 2002. *Natural Selection and Social Theory: Selected Papers of Robert Trivers.* Oxford: Oxford University Press.

Trivers, Robert L. and Dan E. Willard. 1973. "Natural Selection of Parental Ability to Vary the Sex Ratio of Offspring." *Science.* 179: 90–2.

Uller, Tobias and L. Christoffer Johansson. 2003. "Human Mate Choice and the Wedding Ring Effect: Are Married Men More Attractive?" *Human Nature.* 14: 267–76.

US Bureau of the Census. 1995. *Current Population Reports.* Series P60–187. Child Support for Custodial Mothers and Fathers: 1991. Washington, DC: US Government Printing Office.

van den Berghe, Pierre L. 1990. "From the Popocatepetl to the Limpopo." Pp. 410–31 in *Authors of Their Own Lives: Intellectual Autobiographies by Twenty American Sociologists*, edited by Bennett M. Berger. Berkeley: University of California Press.

van den Berghe, P. L. and P. Frost. 1986. "Skin Color Preference, Sexual Dimorphism and Sexual Selection: A Case of Gene Culture Co-Evolution?" *Ethnic and Racial Studies.* 9: 87–113.

Verner, Jared. 1964. "Evolution of Polygamy in the Long-Billed Marsh Wren." *Evolution.* 18: 252–61.

Verner, Jared and Mary F. Willson. 1966. "The Influence of Habitats on Mating Systems of North American Passerine Birds." *Ecology.* 47: 143–7.

Voland, Eckart. 1984. "Human Sex-Ratio Manipulation: Historical Data from a German Parish." *Journal of Human Evolution.* 13: 99–107.

Wagatsuma, Erica and Chris L. Kleinke. 1979. "Ratings of Facial Beauty by Asian-American and Caucasian Females." *Journal of Social Psychology.* 109: 299–300.

Wall, Florence Emeline. 1961. *The Principles and Practice of Beauty Culture*, Fourth Edition. New York: Keystone Publications.

Walter, Tony and Grace Davie. 1998. "The Religiosity of Women in the Modern West." *British Journal of Sociology.* 49: 640–60.

White, Douglas R. 1988. "Rethinking Polygyny: Co-Wives, Codes and Cultural Systems." *Current Anthropology.* 29: 529–58.

White, Gregory L. 1981. "Some Correlates of Romantic Jealousy." *Journal of Personality.* 49: 129–47.

Whitmeyer, Joseph M. 1997. "Endogamy as a Basis for Ethnic Behavior." *Sociological Theory.* 15: 162–78.

Williams, George C. 1966. *Adaptation and Natural Selection: A Critique of Some Current Evolutionary Thought.* Princeton: Princeton University Press.

Wilson, David Sloan. 2002. *Darwin's Cathedral: Evolution, Religion, and the Nature of Society.* Chicago: University of Chicago Press.

Wilson, David Sloan. 2007. *Evolution for Everyone: How Darwin's Theory Can Change the Way We Think about Our Lives.* New York: Delacorte.

Wilson, Edward O. 1998. *Consilience: The Unity of Knowledge.* New York: Knopf.

Wilson, Margo, Martin Daly, and Christine Wright. 1993. "Uxoricide in Canada: Demographic Risk Patterns." *Canadian Journal of Criminology.* 35: 263–91.

Wilson, Margo, Holly Johnson, and Martin Daly. 1995. "Lethal and Nonlethal Violence Against Wives." *Canadian Journal of Criminology.* 37: 331–61.

Wolfgang, Marvin E. 1958. *Patterns in Criminal Homicide.* Philadelphia: University of Pennsylvania Press.

Wong, R. C. and C. N. Ellis. 1984. "Physiologic Skin Changes in Pregnancy." *Journal of the American Academy of Dermatology.* 10: 929–43.

Wright, Robert. 1994. *The Moral Animal: The New Science of Evolutionary Psychology.* New York: Vintage.

Yamagishi, Toshio, Nobuhito Jin, and Toko Kiyonari. 1999. "Bounded Generalized Reciprocity: Ingroup Favoritism and Ingroup Boasting." *Advances in Group Processes.* 16: 161–97.

Yamagishi, Toshio, Shigeru Terai, Toko Kiyonari, and Satoshi Kanazawa. Forthcoming. "The Social Exchange Heuristic: Managing Errors in Social Exchange." *Rationality and Society.*

Yamaguchi, Kazuo and Linda R. Ferguson. 1995. "The Stopping and Spacing of Childbirths and Their Birth-History Predictors: Rational-Choice Theory and Event History Analysis." *American Sociological Review.* 60: 272–98.

Young, Mark C. (Editor.) 1994. *The Guinness Book of Records 1995.* New York: Facts on Life.

# Index

## Index

## Index

## About the Authors

**Alan S. Miller** Until his very untimely death in January 2003 at the age of 44, Alan S. Miller was Professor of Social Psychology in the Department of Behavioral Sciences at Hokkaido University, Japan. He was also Affiliate Associate Professor of Sociology at the University of Washington. He received his BA from UCLA and his PhD from the University of Washington, and had served on the faculties of the University of North Carolina at Charlotte and Florida State University. His last home institution, Hokkaido University, is one of Japan's elite national universities, and Professor Miller was the first non-Japanese academic to be given a permanent, tenured position there. The Department of Behavioral Sciences at Hokkaido University is the leading department in Japan in the area of evolutionary psychology.

Professor Miller was the author of more than twenty-five articles in peer-reviewed academic journals, writing in the areas of crime and deviant behavior, religion, and cross-cultural social psychology. He has written an academic book (with Satoshi Kanazawa) that explores the origin and nature of social order in contemporary Japanese society, *Order by Accident: The Origins and Consequences of Conformity in Contemporary Japan* (Westview, 2000).

**Satoshi Kanazawa** is Reader in Management at the London School of Economics and Political Science, and Honorary Research Fellow in the Department of Psychology, University College London, and in the Department of Psychology, Birkbeck College, University of London. He received his MA from the University of Washington and his PhD from the University of Arizona, both in sociology. He was the first sociologist to introduce modern evolutionary psychology into sociology. His evolutionary psychological work has appeared in peer-reviewed scientific journals in all the major social sciences (sociology, psychology, political science, economics, and anthropology), as well as biology, and he has published more than seventy articles and chapters. He currently serves on the editorial boards of *Evolutionary Psychology*; *Journal of Social, Evolutionary, and Cultural Psychology*; and *Managerial and Decision Economics.* His work has been widely featured in the mass media on several continents, including the *New York Times*, the *Times* (London), the *Washington Post*, the *Los Angeles Times*, the *Boston Globe*, *Time*, *Psychology Today*, the *Times Higher Education Supplement*, the *Times Education Supplement*, the *Australian*, and the *Globe and Mail*, and he has been interviewed on Fox News Live, BBC World Service, BBC Radio 4, NPR's *All Things Considered*, among other TV and radio shows. With Alan S. Miller, he is coauthor of *Order by Accident: The Origins and Consequences of Conformity in Japan* (Westview, 2000).